Motors, Contro
for Air Conditi
and Refrigeratic

Motors, Controls and Circuits for Air Conditioning and Refrigeration Systems

Thomas E. Kissell, MED
Terra Technical College
Fremont, Ohio

Reston Publishing Company, Inc.
A Prentice-Hall Company
Reston, Virginia

Library of Congress Cataloging in Publication Data

Kissell, Thomas.
  Motors, controls and circuits for air conditioning
and refrigeration.

  Includes index.
  1. Air conditioning—Electric equipment. I. Title.
TK4035.A35K57 1984   697.9'32   83-13661
ISBN 0-8359-4665-7

Copyright 1984 by
Reston Publishing Company, Inc.
A Prentice-Hall Company
11480 Sunset Hills Road
Reston, Virginia 22090

10   9   8   7   6   5   4   3   2   1

Printed in the United States of America

Contents

5 Relays and Thermostats 40

6 Basic Hermetic Motor Theory 50

7 The Split-Phase Compressor 62

8 The Capacitor-Start, Induction-Run Compressor 74

9 The Permanent Split Capacitor Compressor 86

Contents

Commercial System Controls 240

Acknowledgments

The author wishes to acknowledge the following people for their support and guidance in the completion of this work: Dr. Paul Brauchle for his help in editing, Dr. Ernest Savage for his help in setting up the outline, and Mr. Anthony Palumbo for his editing of all technical material. My deepest appreciation goes out to these three gentlemen for meeting all deadlines and making the coordination of this project go smoothly.

I would also like to thank my wife Kathy for all the work she put into the typing of this manuscript and for putting up with this project. I also thank my family for their support during this time.

Product Acknowledgments

The author wishes to acknowledge and thank the following companies for their help in locating pictures of their products used in this text. Without their help and co-operation, this book would not be complete.

Allen-Bradley Company
Copeland Corporation
General Electric
Honeywell Incorporated
ITE Electrical Products (a division of Siemens-Allis Inc.)
Ranco Controls

Introduction

The electrical trainer described and referred to in this book is pictured below. It can easily be constructed by mounting the components listed in the parts list on a board. The author used 4' × 4' sheet of heavy peg board mounted on a 2' × 2' × 2' plywood box and attached four casters. This allowed for the storage for all unused parts and wires. It is suggested that any of the parts that must be purchased be quality trade type parts normally found in air conditioning or refrigeration systems. These parts should be readily available at any air conditioning and refrigeration parts wholesaler in your community. It is the author's intention to keep the parts as realistic and "normal" as possible so that the students will be working with parts similar to what they will find when they go out on the job.

Parts List for Trainer

A. 3 pole, 3 phase, 30 amp disconnect
B. 24 volt 40 VA transformer
C. Single-phase disconnect (2 pole, 30 amp)
D. 15 amp, 115 volt duplex
E. Low voltage thermostat with terminal board
F. Defrost timer
G. Low pressure switch
H. High pressure switch
I. Oil pressure switch
J. 3 pole, 15 amp, 24 volt relay
K. 2 pole, 15 amp, 24 volt relay
L. 2 pole, 15 amp, 24 volt relay
M. 2 pole, 15 amp, 120 volt relay
N. 2 pole, 15 amp, 230 volt relay

O. Fan switch (fan/limit combination)
P. Heating sequencer (24 volt coil, 15 amp contacts)
Q. Heating sequencer (24 volt coil, 15 amp contacts)
R. Motor starter (15 amp, 120 volt coil)
S. Start/stop switch
T. Current relay
U. Potential relay
V. Start capacitor for compressor
W. Run capacitor for compressor
X. Small single-phase compressor
Y. Capacitor start motor 115/230 volt, 2 speed
Z. PSC motor 230 volt, 2 speed with run capacitor

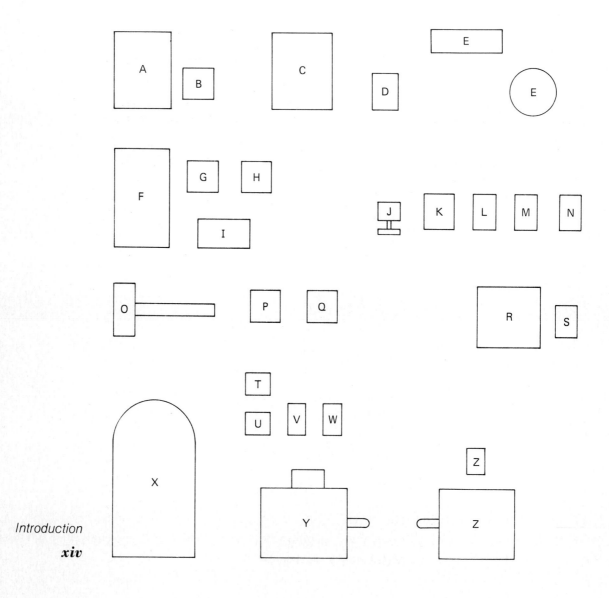

Quizzes

At the end of each lesson a quiz may be given to check each student's mastery of the listed skills. Your instructor will use the objectives listed on the first page of each unit as this quiz. It is important that you practice these skills until you have mastered them before you ask your instructor for the quiz.

Procedures

Units 1 through 3 should be performed in sequence prior to working on any other unit. Units 4 through 6 may be completed in any order. Unit 6 should be completed prior to units 7 through 15. Units 7 through 15 may be completed in any order. Units 16 through 20 should be completed prior to units 21 through 24. Units 16 through 20 may be completed in any order as may units 21 through 24.

Safety

All of the experiments and activities in this book have been successfully field tested. To complete these activities safely you must follow the directions exactly. Any variation in these procedures may cause a serious accident or damage to the equipment.

You must read all the safety rules listed at the beginning of each lesson and strictly follow the instructions provided therein.

The author of this book cannot assume any liability for any changes in procedures or the disregard of the safety rules. Some complex work in the units will request that you get your instructor's okay or approval before you may apply power. Be sure to get this approval and have your instructor inspect your work when requested.

Your instructor may ask you to sign your name at the bottom of each unit's safety rules to insure that you have read the rules and thoroughly understand them. As a general rule of thumb, treat all electrical components and wiring as though they are powered unless you have measured the voltage level with a meter and can prove the power is off.

Motors, Controls and Circuits
for Air Conditioning
and Refrigeration Systems

1 Fundamentals of A.C. Electricity

Safety for Unit 1	Tools Required
At times the circuits and disconnects you will be working with will be powered with up to 230 volts. You must be aware of and take appropriate safety precautions relative to electrical shock hazards and wear safety glasses while working.	Voltmeter (0-250 volt scale) Screwdriver (flat blade ¼ in.)

Objectives for Unit 1

At the conclusion of this unit the student should be able:

To explain the following terms either orally or in written form to the instructor's satisfaction:

a) Alternating current

b) Frequency

c) RMS or effective voltage

To explain verbally or in writing the difference between neutral and ground in an A.C. circuit to the instructor's satisfaction.

To use a sine wave to explain the operation of a simple single-phase generator to the instructor's satisfaction.

To use a VOM to measure accurately the following in the single-phase disconnect:

a) Voltage from line 1 or line 2 to ground.

b) Voltage from line 1 or line 2 to neutral.

c) Voltage from line 1 to line 2.

d) Which parts of the circuit are grounded.

To indicate the colors for ground wires and neutral wires as specified by the National Electric Code.

To identify accurately the parts of a single phase disconnect box.

Before starting Unit 1 the student should be able to use a volt-ohmmeter (vom) to measure voltage and resistance in an electrical circuit. The student should also have successfully completed a basic electricity course and understand the terms ohms, volts, amps, magnetism, induction, and flux lines.

MEASURING VOLTAGE IN THE DISCONNECT BOX

As an air conditioning technician, you will be required to locate the source of electricity on the job and install conductors to provide a circuit to the air conditioning equipment. The normal source will be a *disconnect box* (see figure 1-1 below). The disconnect box is an enclosure that contains the terminal connections for the conductors and a switch that allows the power to be turned off while work on the system is in progress. During the installation process you will be required to open the box and make all necessary connections.

Locate the disconnect box on your trainer. Be careful to turn the handle to the OFF position before trying to open the box. This is accomplished by pulling down the handle on the right side of the box.

It is important to remember that, even though the disconnect is turned off, there will still be power at the top terminals inside of the box (see Figure 1-1). This is called the *line side* of the disconnect.

A voltmeter should be used to check the *line side* to see if it is powered. Set the voltmeter range selector to the 250 volt A.C. range. Place the meter probes on the line side terminals as shown in figure 1-2.

RECORD THE VOLTAGE L_1 to L_2 _____

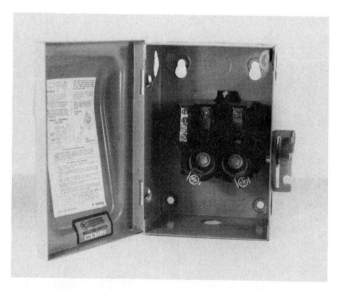

Figure 1-1 Disconnect box (courtesy of *ITE Electrical Products*, a division of Siemens-Allis, Inc.)

Place one probe on line 1 terminal and one probe on the neutral bar as shown in figure 1-3.

RECORD THE VOLTAGE L_1 to N _____

Place one probe on L_2 terminal and one probe on the neutral bar.

RECORD THE VOLTAGE L_2 to N _____

The voltage you have measured in the disconnect is of course not produced by the disconnect, but instead has traveled quite some distance to get to the disconnect. The source of the current is an electrical generating station operated by an electric utility company. The utility company uses a fuel such as coal, fuel oil, or nuclear power to make steam which drives a turbine. The turbine turns a generator to produce the current.

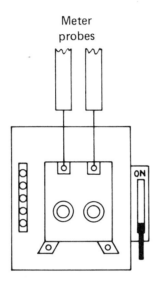

Figure 1-2

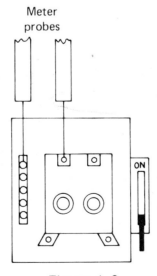

Figure 1-3

The generator produces the current by a moving coil of wire which passes through a magnetic field. The electrons in the conductor of the coil are induced to flow when the conductor cuts across the flux lines of the magnet. The number of electrons flowing will increase as more flux lines are cut. The coil in the generator is on a shaft and moves through 360° (a full circle) during every rotation of the shaft.

If we plotted the output on a graph to show how the current is generated as the coil cuts through the flux lines of the magnet, we would obtain a double curved line. This line would look something like a backward "S" lying on its side. Such a graph is called a waveform. The waveform of the induced current will look like the waveform in figure 1-4.

Since the waveform goes to the peak positive and peak negative, the current is said to alternate. This is where the name *alternating current*, or A.C., comes from.

The voltmeter can be used to show this. Open the disconnect box and put the red probe on L_1 and the black probe on neutral.

RECORD THE VOLTAGE _____

Now reverse the probes by putting the red probe on the neutral and the black probe on L_1.

RECORD THE VOLTAGE _____

You should notice the meter needle deflected in the same direction both times. This has occurred because the voltage is alternating too.

As an air conditioning technician you will be required to provide explanations of equipment operation, efficiency, or problems to customers. In order to do this you must understand the fundamentals of electricity yourself. You must be able to apply electrical degrees to the sine wave, and understand such terms as frequency, RMS, and effective voltage. The diagram of figure 1-5 is provided for you to add the degrees of the shaft rotation to the sine wave.

Start by adding zero to the point where the wave begins. There is no current output at 0°. The wave goes positive to the peak. This indicates maximum current output. The generator shaft has moved one-fourth turn, or 90°. Add 90° to this point. The current returns to zero during the next one-fourth turn. The generator shaft has now moved a total of 180°. Add 180° to the waveform at this point. Now the current moves in the opposite direction where it peaks negative. The point where the current returns to zero

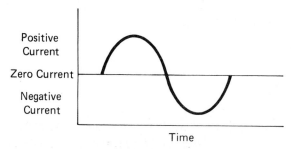

Figure 1-4

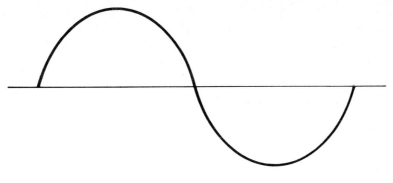

Figure 1-5

again is the point at which the shaft has completed a full cycle (360°). Mark 360° on your sine wave.

BASIC DIAGRAM OF POWER SOURCE AND DISCONNECT

Since your installation work as an air conditioning and refrigeration technician will start with determining conductor terminal connections in the disconnect box, more information about the disconnect is needed. The 230 or 208 volts that you measure on L_1 and L_2 (line 1 and line 2) in the disconnect comes from the secondary of a transformer. (The operation of transformers will be discussed in Unit 2.) The diagram looks like this (see figure 1-6).

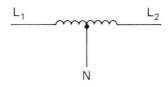

Figure 1-6

When the disconnect box is added to this diagram the transformer coil is not shown, only its terminals L_1, L_2, and neutral. An example of a power supply and disconnect is shown in figure 1-7.

A.C. TERMS

Each sine wave is considered to be one cycle. Frequency is the number of cycles that occur in one second. The frequency of the electricity supplied to industrial and residential customers is 60 cycles per second. The unit of

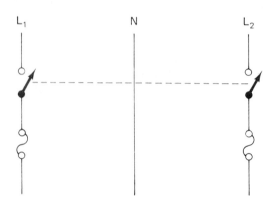

Figure 1-7

measurement of frequency is Hertz. Therefore the supply voltage frequency of 60 cycles per second is often said to be 60 Hertz.

Another important term for A.C. is RMS. RMS stands for root mean square. It is really a mathematical conversion to show what amount of A.C. produces the same heating effect as an equal voltage of direct current. This conversion is necessary since the output of the alternating current is always changing (remember the sine wave). Since the current is sometimes peak and sometimes zero, a mathematical conversion is needed to show the value equal to D.C. For simplicity all A.C. voltages are listed in their RMS values, unless otherwise stated. For instance the 120 volts you previously read with your meter is RMS volts. The RMS term is usually dropped or taken for granted.

NEUTRAL AND GROUND

Neutral is the midpoint between terminals L_1 and L_2 on the transformer (refer to figure 1-6). The voltage from either L_1 or L_2 to neutral will be 115 or 120 volts. Treat L_1 and neutral or L_2 and neutral as you would the + and − of a D.C. power supply. The neutral will be the return path for all 115 or 120 volt currents.

The National Electric Code (NEC) specifies that the neutral wire be white or light gray in color. Locate the neutral bar in your disconnect. RECORD THE COLOR OF YOUR NEUTRAL WIRE _____

All air conditioning and refrigeration equipment must be grounded. The symbol for ground is the same as in D.C. circuits: ⏚ This means that all metal cabinets, frames, and housings in the equipment must be connected to ground. This can be accomplished through conductors or through the metal of the frame, cabinet, or metal tubing.

Grounding will cause any current-carrying conductor to "short out" or draw excessive current if it comes in contact with the frame or cabinet. The excessive current draw should cause the circuit fuse to open, interrupting current flow.

The components on your trainer should be grounded, just as in the field. In most circuits the ground is connected to the neutral only in the first disconnect (the main). In all other disconnects the neutral bar is not grounded, to prevent circulating currents through the ground wires. Every disconnect's metal case, however, must always be grounded.

The electrical system is grounded to make the frames and neutral the same potential as the earth ground. You will use an ohmmeter to test a system to determine if the neutral and frame are grounded.

Turn the disconnect switch off. Be sure this has been accomplished by testing for voltage at the load-side terminals of your disconnect. If any voltage is present, double-check the disconnect or call your instructor. When you are sure there is no voltage, set the meter to the lowest ohms scale and zero the meter. Now place one probe on the neutral bar and touch the side of the disconnect box. RECORD YOUR OHM READING _____ If your reading is not zero your neutral has become ungrounded. Call your instructor to ground the neutral bar.

It is important to be sure all the components on your trainer are also grounded. To do this, leave one lead of the ohmmeter on the ground point (neutral bar) in the disconnect box and touch the components. These parts should include the housing of the disconnect and interconnecting electrical tubing. Your ohmmeter should read zero ohms for every test. If the meter reads infinity (∞) while touching any metal part, that part is not grounded and you should notify your instructor. Remember: by grounding all metal parts around the electrical system you are ensuring that any time a line wire touches a metal frame the circuit fuse will open. This allows you to touch the metal parts of the system without worrying about being shocked.

All ground wires will be green or bare (without an insulating coating). Locate the ground wires in the disconnect. RECORD THEIR COLORS

_____

Remember, the neutral wire will carry current in any 115 or 120 volt circuit. The ground will carry current only during a short circuit or fault.

You now have completed Unit 1. Take this time to review the objectives listed at the beginning of this unit. If you feel you can successfully complete each objective, call your instructor for a short quiz on the unit objectives. Your instructor will determine whether or not you have sufficient competency to go on to Unit 2.

2 Three-Phase Power Distribution

Tools Required

Voltmeter (0-250 volt scale)

Screwdriver (flat blade ¼ in.)

Objectives for Unit 2

At the conclusion of this lesson the student should be able:

To draw the degrees in their proper locations on a sine wave diagram to the instructor's satisfaction.

To explain orally or in writing why a three-phase voltage system does not need a neutral wire.

To measure accurately the voltage from a three phase disconnect and determine if the voltage comes from a delta or a wye transformer.

To use a voltmeter to determine the loss of a phase and find the open fuse in a three phase disconnect.

To draw a diagram of a delta wired transformer and indicate the voltage from L_1 to L_2, L_2 to L_3, and L_3 to L_1.

To draw a diagram of a wye wired transformer and indicate the voltage from L_1 to L_2, L_2 to L_3, and L_3 to L_1.

THE NATURE OF THREE PHASE VOLTAGE

Air conditioning and refrigeration equipment installed in commercial and industrial buildings may require three phase voltage. Three phase voltage (usually indicated by the symbol 3φ voltage) is generated as three separate A.C. sine waves (see figure 2-1.) If three phase voltage is present on the site, it is usually preferred to power the equipment. This is because 3φ equipment is more efficient than single phase equipment.

The on-site source for 3φ voltage will be a 3φ disconnect (see figure 2-2). The 3φ disconnect can be identified by the presence of three fuse holders in the disconnect. To determine if the disconnect switch has 3φ voltage at its line-side terminals, complete the steps in the following procedure. Check off the steps as you complete them.

1. Locate the 3φ disconnect on your trainer.

2. Turn the disconnect switch off by pushing down on the handle where it is marked "off." (Remember the line-side terminals at the top of the box will still be powered.)

3. Open the door to the disconnect switch by pulling out on the top right corner of the door.

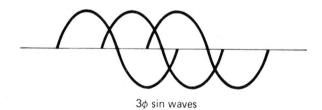

3φ sin waves

Figure 2-1   30 or 3 phase sine waves

4. Inspect the disconnect. Notice there are three fuses, three line side terminals, and three load side terminals. The terminals and fuses are identified from left to right as you look in the box. They are called line 1, line 2 and line 3. Sometimes they are also called L_1, L_2, and L_3, or phase A, phase B, and phase C.

5. Set your voltmeter at 250 volts and measure the 3φ voltage.

    a. Place the probes of your meter on the top left and top center terminal and record the voltage. This will be the voltage from line 1 to line 2. L_1 to L_2 voltage _____

Figure 2-2 (Courtesy of *ITE Electrical Products*, a division of Siemens-Allis, Inc.)

b. Place the probes on the top middle terminal and the top right terminal and record the voltage. This will be the voltage from line 2 to line 3. L_2 to L_3 voltage _____

c. Place the probes on the top left terminal and top right terminal and record the voltage. This will be the voltage from line 3 to line 1. L_3 to L_1 voltage _____

d. If each of the readings in step 5a, 5b, and 5c gives the same amount of voltage, the disconnect is powered with 3ϕ voltage.
Turn the disconnect off and close the door. Continue on to the next section.

Three phase voltage comes from the electric utility generator as three separate sine waves. Remember from Unit 1 that a single sine wave has 360° and is produced by one coil on the generator shaft. In three phase voltage the three sine waves are produced from three separate coils placed on the generator shaft 120° apart. This produces the three sine waves 120° apart.

To get a better understanding of the three sine waves that makeup three phase voltage, refer figure 2-3. Observe that three sine waves, L_1, L_2, and L_3, are plotted. Remember that the three sine waves are 120° apart,

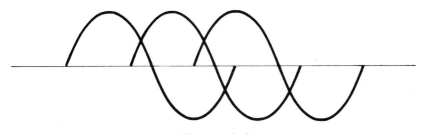

Figure 2-3

The Nature of Three Phase Voltage

13

and that one sine wave has 360°. On each of the sine waves below indicate on the diagram the degrees (0, 90, 180, 270, and 360) at their proper positions.

In a three-phase voltage system the three generated voltages are supplied to the disconnect through three separate wires. These conductors also provide the return path for the current. This means that current can flow to the load through L_1 and return through L_2 at one instant and be supplied through L_3 and return through L_1 the next instant. The exact routing is determined by the alignment of the coils on the shaft of the three phase generator. The point to remember here is that a three phase system does not require a neutral wire to provide a return path for current, because there is always a line available for return voltage.

The three phase system does require a *ground* for all metal cases and cabinets. The ground is of course required for all electrical circuits. For the three phase system the ground wire will be green or bare as in the single phase system.

The voltage is distributed from the generating station through transmission lines at very high voltage. When the voltage arrives at a substation near the industrial site, a set of transformers lowers or steps it down to 4,160 volts. The 4,160 three-phase is distributed around town to specific commercial and industrial areas. At the industrial site the voltage is again stepped down by transformers to an appropriate level, usually 230 or 208 volts (see figure 2-4 for the transmission system).

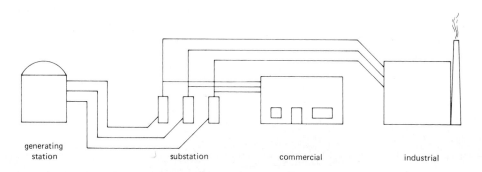

generating
station   substation   commercial   industrial

Figure 2-4

For the transformer to provide 208 volts the three transformer windings will be wired in a wye configuration (see figure 2-5).

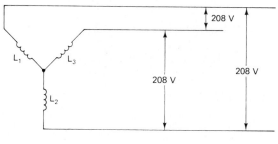

Figure 2-5

This means that between L_1 and L_2 or L_2 and L_3 the voltage will measure 208 volts. It also means that the voltage between any two of the three lines will measure 208 volts.

To provide 230 volts to the customer the transformer windings must be wired in a delta configuration (see figure 2-6).

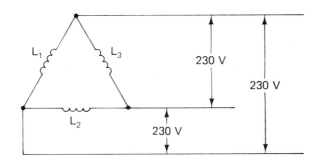

Figure 2-6

Now the voltage between any two lines will measure 230 volts.

Make the voltage measurements in your 3ϕ disconnect again (follow the procedure on pages 1 and 2). This time indicate whether the transformer supplying the voltage to your disconnect is wired in delta or wye configuration. Amount of voltage _____. Type of configuration _____.

On some job sites the voltage source will be a three phase load center or circuit breaker box instead of a disconnect (see figure 2-7).

Figure 2-7   Load center (courtesy of *ITE Electrical Products*, a division of Siemens-Allis, Inc.)

Some air conditioning and refrigeration manufacturers require that a fused disconnect be installed in addition to the load center. All air conditioning and refrigeration equipment is required by the National Electric Code to have a disconnect switch installed within 10 feet of the equipment. Having the disconnect close to the equipment serves two purposes: it allows the technician to shut off the power when work is required on the equipment, and it is a convenient place to locate fuses for the equipment.

The disconnect on your trainer is the fusable type. The switch allows you to turn off power for some work procedures, and the fuses protect the trainer's components. There will be more about fuses in future units.

The three phase disconnect has power brought in on the line-side terminals and taken out off the load-side terminals. The switch always opens power between the line side and the fuse. This permits the fuses to be safely removed when no power is applied.

Use the following steps to demonstrate that your disconnect will not be powered at the fuse clip when the disconnect is turned off. (Check the steps off as you complete them.)

1. Set your voltmeter to 250 volts.

2. Open the disconnect door. Place the meter probes on the line side terminals of L_1 and L_2.

3. Turn the disconnect on and record the voltage _____.

4. Turn the disconnect off and record the voltage _____.
(Notice the line-side terminals L_1 and L_2 are "powered" when the switch is in the OFF position as well as in the ON position.)

5. Now place the probes on the top fuse clip of L_1 and L_2.

6. Turn the disconnect on and record the voltage _____.

7. Turn the disconnect off and record the voltage _____.
(Notice there is no voltage on the top fuse clip of L_1 and L_2 when the disconnect is turned off. The voltage is not present on the top fuse clip because the disconnect switch has interrupted the voltage. This allows you to change the fuse or work on the load-side terminals with power off.)

8. Turn the disconnect off and close the door.

CHECKING THREE PHASE VOLTAGE

Another important job for the air conditioning technician is to determine the condition of a three-phase voltage supply. If an air conditioning or refrigeration system fails to operate properly, the problem may be with the voltage in the three phase supply. Frequently one or more of the fuses in the load center or disconnect will open. This will cause the loss of one phase, and lead to a condition known as *single phasing*. Even though two lines still provide voltage, only one circuit path is available for current.

Use the following procedure to check the disconnect for a bad fuse.

1. *Ask your instructor to insert one bad fuse.*

2. Set your voltmeter to 250 volts range.

3. Open the door on the disconnect.

4. Turn the disconnect on.

5. Place your probes on the load side of the following terminals and record the voltages.

   a. L_1 to L_2 _____

   b. L_2 to L_3 _____

   c. L_3 to L_1 _____

6. If one of the fuses is blown, only one of the readings will show voltage. The two lines that are used when the meter measures full voltage have good fuses. The line that is not being used for that reading will have the bad fuse.

   Remove the bad fuse and test for continuity with your ohmmeter.

   Record the meter reading _____

   a. If the fuse you have just checked shows a reading of zero (0) ohms you have made a mistake since the fuse is good. Ask your instructor to watch you repeat the test procedure to find your mistake.

   b. If the fuse you have just checked shows a reading of infinity (∞) on the highest meter scale, you have found the bad fuse. You may ask the instructor to insert the bad fuse in a different circuit if you would like to practice this procedure again. Repeat the process starting with step 1.

   If you feel you can find the bad fuse on every test, continue on to the next procedure.

FINDING TWO BAD FUSES

Use the following procedure to test for two bad fuses.

1. Ask the instructor to insert two bad fuses into your disconnect.

2. Turn the disconnect on and make the voltage measurements on the following load side terminals:

   L_1 to L_2 _____

   L_2 to L_3 _____

   L_3 to L_1 _____

   (Notice that none of the measurements is the full applied voltage. These readings will be zero or some small voltage.)

3. Now test the voltage from each line side terminal to ground.

L₁ to G _____

L₂ to G _____

L₃ to G _____

4. The line that shows voltage has a good fuse. The lines that show zero have the bad fuses. Remove the bad fuses from the disconnect and check for continuity.

a. If the continuity test shows that either of the fuses is good, call your instructor to watch you repeat the steps of this procedure and make corrections where required.

b. If the continuity tests show both the fuses are bad, you have successfully completed this procedure.

5. Turn off the disconnect and close the door.

Some technicians prefer to remove all three fuses and test them for continuity immediately. This procedure takes a little more time than the voltage tests, and it is not so complete, since it is possible that the problem is with the power supply closer to the transformer. The voltage check would indicate this. However, either test is accurate and acceptable.

You have now completed Unit 2. Review the objectives at the beginning of this lesson. Practice ones you have trouble completing. Call your instructor when you feel you are ready to be quizzed on the objectives.

3 Single Phase Power Distribution

Tools Required

Wire stripping tool (for #10 AWG through #18 AWG wire)

Screwdriver (¼ in. flat blade)

Voltmeter (250 volt scale)

Objectives for Unit 3

At the end of this lesson the student should be able:

To wire correctly and safely a single-phase disconnect switch (two-wire voltage system) from a three-phase disconnect switch.

To wire correctly a single phase disconnect switch (3 wire $L_1 - L_2 - N$ voltage system) from a three-phase disconnect switch.

To measure the voltage in a single phase disconnect to within 10% of actual voltage (L_1 to L_2, L_1 to N, and L_2 to N).

To explain in writing or orally to the instructor's satisfaction where a high leg delta or wild leg would be found.

To draw a ladder diagram of a single-phase voltage system derived from a three-phase disconnect switch to the instructor's criteria.

To install correctly and safely a duplex receptacle and operate a 110-volt power tool or 110-volt lamp from the receptacle.

TYPES OF SINGLE-PHASE POWER SYSTEMS

In some applications single-phase air conditioning equipment is specified even though three phase power is available. The air conditioning technician must be able to identify and use the proper terminals if a three phase system is used to supply power to the single phase disconnect. There are basically two types of single-phase power systems used in air conditioning systems.

The most common system uses a two wire system. In the two wire system, voltage can be 208 or 230 volts. If the voltage is 230 volts, the utility company connects the three transformer secondary windings in the shape of a triangle (see figure 3-1). This configuration is called a *delta*. If the voltage is 208 volts the utility company connects the three transformer secondary windings in the shape of the letter "Y." This configuration is called a *wye* (see figure 3-2).

The two wire system works by having current come in L_1 to the load and return through L_2 for the first half cycle. Then the current reverses and comes in L_2 to the load and returns via L_1. Since only two lines are needed for this system, they could come from either L_1 and L_2, or L_2 and L_3, or L_3 and L_1. The second type of power system will be discussed later.

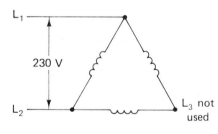

Figure 3–1

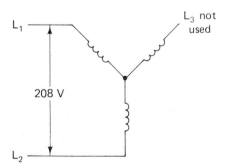

Figure 3–2

WIRING A SINGLE PHASE DISCONNECT

Locate the single phase disconnect on your trainer. (Refer to Unit 1 if you need help identifying the single phase disconnect.) In this unit you will select and run the wires from the three phase disconnect to the single phase disconnect. Use the following procedure to wire the single phase disconnect. Check off the steps as you complete them.

1. Use your voltmeter to be sure the three phase disconnect is not powered. For this activity make sure your trainer is unplugged. Open the three phase disconnect and measure the voltage from L_1 to L_2, L_2 to L_3, and L_3 to L_1.
   Record your voltage _____.

2. If the voltage in step 1 is not zero, call your instructor before proceeding. If voltage is zero proceed to step 3.

3. Select two AWG #12 wires from your wire supply. Measure and cut the wire to make the connection between the load side terminal of the three phase disconnect and the line side terminal of the single phase disconnect. Leave about six inches extra on each end to work with in the disconnect. According to the National Electric Code (NEC), these line wires can be any color except green, white, light gray, or bare.

4. Use a wire stripping tool to strip back the insulation on each end of wire so that approximately ½ in. of stranded conductor is showing (see figure 3-3). Inspect each end to assure that none of the strands are missing.

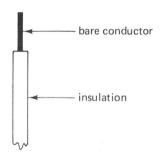

bare conductor

insulation

Figure 3-3

If you have cut any of the strands you must cut the wire back ¾ in. and strip the end again. Note: cut strands reduce the current carrying capacity of the wire.

5. Open the doors of the three phase and single phase disconnects. Insert the two wires through the conduit (metal tubing) that connects the two boxes. A ground wire is not needed because the metal tubing will act as the ground between the two disconnects.

6. Locate any two of the three *load side* terminals in the three phase disconnect. Loosen the screws on these terminals and slide the bare end of your wire under the terminals and tighten the screws securing the wires (see figure 3-4).

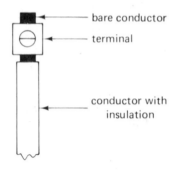

bare conductor

terminal

conductor with
insulation

Figure 3–4

Notice in the diagram that only approximately 1/16 in. of bare copper extends through the top of the terminal and that the insulation nearly touches the bottom of the terminal. This is a necessary safety precaution. If too much bare wire is left above or below the terminal, it may touch other conductors and cause a short circuit.

An important point should be made now. Nearly every electrical installation that you make will be inspected by a state or local inspector. The inspector will form a very critical first impression of the quality of your work by examining the terminal connections in your disconnect. If they are sloppy and haphazard, the inspector will have the impression that you do not have proper skills, and the quality of your work will always be suspect.

7. Loosen the two line-side terminals of the single phase disconnect and make connections there. Again be aware of the quality of your work.

8. Check the diagram (figure 3-5) to be sure your connections are correct. Be sure to obtain your instructor's approval prior to proceeding to step 9.

   Instructor's approval _____

9. Plug the trainer into the receptable to power the three phase disconnect. Turn the three phase disconnect on and measure the voltage at the load-side terminals L_1 to L_2 _____.
   If you do not have voltage L_1 to L_2 call your instructor.

10. Measure the voltage at the line-side terminals of the single phase disconnect L_1 to L_2 _____. Remember if your voltage is 208 volts it comes from wye wired transformers; if it is 230 volts it comes from delta wired transformers.

    a. If you have voltage L_1 to L_2 at the line side of the single phase disconnect, you have successfully completed this procedure. Proceed to the next part of this unit.

    b. If you do not have voltage at the line-side terminals L_1 to L_2 of the single phase disconnect, recheck the voltage in the three phase disconnect. If there is voltage at the three phase disconnect but not at the single phase disconnect, you have made a wiring error. Call your instructor to help you check your work.

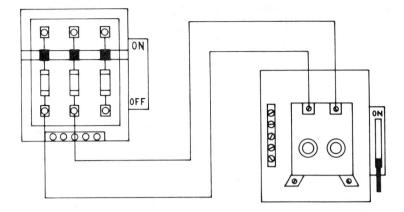

Figure 3-5

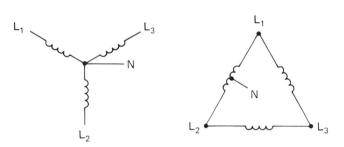

Figure 3-6

WIRE SINGLE PHASE SYSTEM

The second type of single phase system is a three wire system. This system is used where 208 or 230 volts single phase and 115 volts are both needed. An example of this system is your home. 230 volts is needed for the air conditioning system and 115 volts is needed for the gas furnace. To get the 115 volts, the single phase transformers must be tapped for neutral. Neutral is the midpoint between L_1 and L_2 of the transformer secondary. Neutral for the wye and delta systems is shown below (see figure 3-6).

The ladder diagram for the delta and wye system is listed below (see figure 3-7).

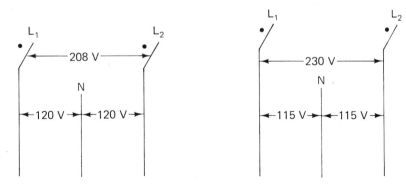

Figure 3-7

The voltages that you would find between each line and neutral are listed on the diagram.

WIRING THE THREE-WIRE SINGE-PHASE SYSTEM

Use this procedure to install the wires for a three-wire single-phase voltage system. Be sure to unplug the trainer any time you are making connections in the disconnect box.

1. Check the voltage in your three phase disconnect to be sure your power is off. Record the voltage L_1 to L_2 _____ L_2 to L_3 _____ and L_3 to L_1_____.

2. If any voltage is present call your instructor. When the voltage is zero proceed to step 3.

3. Remove the two wires you installed between the disconnects for the two wire system. These wires will be reused, so be careful not to damage them.

4. Select a white wire for the neutral of the three wire system. Measure and cut the wire to length as described in the two wire system. This time the wire will have to reach from the neutral bar in the three phase disconnect to the neutral bar in the single phase disconnect.

5. Install the two line wires with the neutral wire through the conduit from the three phase disconnect to the single phase disconnect. Make the terminal connection in each box. Use L_1, L_2, and neutral terminals in each box.

6. Have your instructor check your work before you apply power to the system.
   Instructor's approval _____

7. Apply power to your trainer and turn on the three phase disconnect. Measure and record the voltage at the load-side terminals of the three phase disconnect L_1 to L_2 _____, L_1 to N _____, L_2 to N _____.

8. Measure and record the voltage at the line side terminals of the single phase disconnect. L_1 to L_2 _____, L_1 to N _____, L_2 to N _____. If your voltage L_1 to L_2 is 208 and L_1 to N is 120 (230 L_1 to L_2 and 115 L_1 to N for delta system), you have successfully wired a three-wire single-phase system from a three phase system.
   Remember, L_3 could be used instead of L_2. If time permits you may repeat the procedure using L_1 and L_3 instead of L_1 and L_2. The resulting voltages should be the same.

THE HIGH LEG DELTA SYSTEM

When using the three wire system in industrial areas, you must make one extra check if the voltage is 230 volts. Some three-phase delta wired transformers have one of the secondary windings tapped midway to produce the neutral (see figure 3-8).

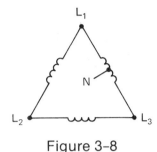

Figure 3–8

The center tap produces 115 volts when used as a neutral with L_1 or L_3. The problem is created between L_2 and N. The voltage between L_2 and N will be 208 volts. This voltage *is not usable* since it causes unequal loading of the transformer. This 208 voltage is called a *high leg delta*. Line 2 is called the "wild leg" and should be identified with an orange wire. Using L_2 to N for 115 volt tools or motors will cause damage due to excessive voltage.

It is important to remember these facts about the high leg delta system. It will only be found in industrial or commercial areas. It will never be used in residential areas. L_2 to N produces 208 volts—which is unusable. L_2 used with L_1 or L_3 will produce 230 volts—which is usable.

Your instructor will locate a high leg delta voltage system in your building (if available). Under supervision of your instructor, make the following measurements in the disconnect switch on a high leg delta system:

L_1 to L_2 _____ L_2 to L_3 _____ L_3 to L_1 _____

L_1 to N _____ L_2 to N _____ L_3 to N _____

Check your measurements using the information given above.

WIRING A DUPLEX RECEPTACLE

On some jobs you will be required to install and wire a duplex receptacle (see figure 3-9) to provide a place to plug in service equipment such as vacuum pumps, inspection lights, and electrical power tools that will be used on the job.

The duplex will be wired through the single phase disconnect L_1 to N to provide 120 volts on a wye system or 115 volts on a delta system. Locate the duplex receptacle on your trainer. It will be mounted in a 2 in. × 4 in. metal box. Remove the duplex from its box so you can get a better look at it. Notice that the duplex has silver, gold, and green screws for connections. The line wire will be connected to the gold screws, the neutral will be connected to the silver screw, and the ground wire will be connected to the green screw.

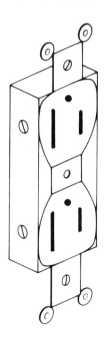

Figure 3-9   Duplex receptacle

Use the following procedure to install a duplex receptacle on your trainer. Check off the steps as you complete them.

1.   Select a line "hot" wire, neutral wire, and ground wire to connect the duplex receptacle to the load side of the single phase disconnect. Cut these wires to the proper lengths. Be sure your wires are the proper color.

2.   Prepare the ends of your wires with the stripping tool.

3.   Install the three wires into the conduit leading from the single phase disconnect to the duplex electrical box.

4.   Loosen the gold screw. Bend the stripped end of your line wire in the shape of a hook and place the wire around the gold screw in a clockwise direction and tighten (see figure 3-10). By placing the wire in the clockwise direction the screw will tend to tighten the wire around it.

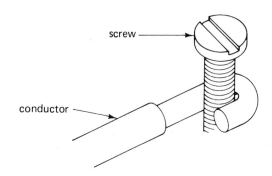

screw

conductor

Figure 3-10

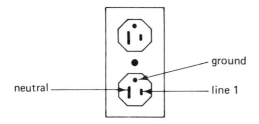

neutral

ground

line 1

Figure 3-11

5. Repeat step 4 for the neutral and ground wires. Use the silver screw for the neutral and the green screw for the ground.

6. Connect the line, neutral and ground wires to the L_1, N, and ground terminals respectively in the single phase disconnect.

7. Have the instructor approve your wiring before you mount the duplex back into its electrical box.

   Instructor's approval _____

8. Mount the duplex in its box and replace its cover.

9. Turn the disconnect on and measure the voltage at the duplex L_1 to N _____ and L_1 to ground _____ (see figure 3-11 for location of L_1, N, and G).

10. If your voltage is between 115 and 120 volts you are ready to plug a power tool or lamp into the duplex and test it. The instructor will provide the lamp or power tool.

    Note: If you do not have the proper voltage at the duplex receptacle, call your instructor.

11. If the lamp or power tool works correctly when you power it from your duplex, you have successfully completed this procedure. Turn off the power and remove the wire from the single phase disconnect and the duplex. Store the wires under your trainer for future use.

   You have now completed all work for Unit 3. Review the objectives to be certain that you can successfully complete them. Call your instructor for the end-of-unit quiz when you are ready.

4 The Control Transformer

Tools Required

VOM
Screwdriver (¼ in. flat blade)

Objectives for Unit 4

At the conclusion of this unit the student should be able:

To identify the control transformer by selecting the correct component on the trainer.

To install a control transformer and connect its primary to proper voltage. This objective will be considered completed when the secondary voltage measures 24 volts ± 5%.

To explain the simple transformer theory orally or in writing to the instructor's satisfaction.

To identify properly when given a control transformer the primary and secondary coils.

To identify accurately when given a control transformer and a single phase power supply the input voltage necessary to obtain 24 volts from the secondary coil.

To calculate the maximum secondary current from the control transformer's volt ampere rating within ± 5%.

To draw properly the control transformer in ladder diagram and wiring diagram forms to the instructor's satisfaction.

To use proper troubleshooting procedures to locate problems accurately in the control voltage circuit.

CONTROL TRANSFORMERS

Most residential air conditioning systems use 24 volts to power the control circuits. Since the residential voltage is usually 230 volts when measured line-to-line and 115 volts when measured line-to-neutral, the equipment manufacturer will usually supply a control transformer. The control transformer (sometimes called the low voltage transformer) will change the line voltage from 230 or 115 volts to 24 volts. The control transformer used in air conditioning systems and its electrical symbol are shown in figure 4-1.

The National Electrical Code classifies transformers according to output voltages. Since the control transformer's voltage is less than 50 volts, it is classified as a Class II transformer.

Locate the control transformer on your trainer. Notice that it has two wires leading into one side and two screw terminals on the other side. The wires are connected to a coil of wire called the transformer primary. The primary side of the transformer is where the voltage comes into the transformer.

The two screws are connected to another coil. This is the point where the 24 volts comes out and it is called the transformer secondary. The secondary terminals are identified by the letters R and C. The C is for

Primary

Secondary

R   C

A   B

Figure 4-1   (a) (Courtesy of *Honeywell Inc.*); (b) electrical symbol of control transformer

common and the R is for red, which is usually the color of wire connected to the terminal. During installation and on diagrams, these terminals will be referred to as R and C.

INSTALLATION OF CONTROL TRANSFORMERS

To install the control transformer follow the steps outlined here. This checklist should be used every time you install a control transformer. Be sure to check off the steps as they are completed.

1. Determine the primary voltage from the data plate on the transformer and record: _____

2. Determine the load-side terminals in the single phase disconnect that will supply the primary voltage. (Refer to Unit 3 if you need help.) Open the disconnect and measure the voltage at the terminals you have selected.
   Record the measured voltage _____
   Be sure it matches the transformer's requirements.

3. Turn off the power to the trainer by unplugging it. Use the voltmeter to be sure there is no voltage at the single-phase-disconnect line side terminals. Measure and record the voltage at the line side terminals _____

4. Make sure the transformer is securely mounted. (If your transformer is already mounted you may skip this step. In the field you would use this step to remove the inoperative transformer and install the new one.)

5. Connect the primary wires to the proper disconnect terminals and tighten the screws. Be sure to observe the color code if you are using the 120 volt primary (black to L_1, white to neutral). Check your wiring against the drawing of figure 4-2.

6. Plug the trainer power supply in and turn the single phase disconnect on.

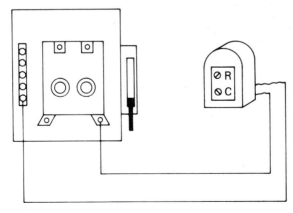

Figure 4-2   Transformer wired to the disconnect box

7. Check the transformer secondary voltage and record _____. (This voltage may be slightly higher than 24 volts but never lower. If your voltage is lower than 24 volts, review steps 1 through 6.)

8. Installation is now complete. Turn power off and remove the transformer primary wire from the single phase disconnect.

PRIMARY AND SECONDARY WINDINGS

The control transformer contains two separate coils. One is called the primary and the other is called the secondary. These coils are made by looping a long wire around an iron core (see Figure 4-3). Each coil is isolated from the other, which means they are not electrically connected. Use your ohmmeter to prove this. Set the ohmmeter on R $\times$ 10 and zero. Now test the primary coil for continuity. The meter should show some resistance. If the reading on R $\times$ 10 scale is infinity (∞), use the highest resistance scale and repeat the measurement. If the meter reads infinity (∞) on the highest scale, the winding has developed an open and the transformer must be replaced. Record your resistance _____.

Repeat this process on the secondary winding. Record the secondary resistance _____. To show that the primary coil and secondary coil are not connected, test one primary lead to one secondary terminal. Your measurement should be infinity (∞) on the highest ohm scale.

Repeat this process using any combination of one primary wire and one secondary terminal. All readings should be infinite (∞), proving there is no electrical connection between primary and secondary coils.

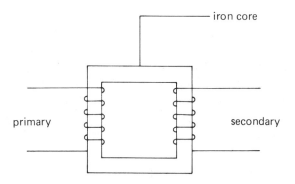

Figure 4-3   Basic transformer

Some transformers have three primary wires and two secondary wires. The extra wire allows the transformer primary to be wired to either 208 or 230 volts and still have 24 volts at the secondary. The diagram is shown in figure 4-4.

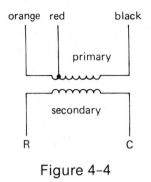

Figure 4–4

THEORY OF OPERATION

To understand how the transformer changes the line voltage into 24 volts, you need to know about the *turns ratio* between the primary coil and secondary coil. This is the number of turns of wire in the primary coil compared to the number of turns in the secondary coil. In the control transformer there will be more turns of wire in the primary coil than in the secondary.

The basic principle of operation of a transformer is induction. It can be explained like this: As voltage is applied to the primary coil, current begins to flow in the primary loop. The current causes a magnetic field to build up sending out flux lines. As the current alternates, as in the sine wave, it builds to a peak and returns to zero. This action causes the flux field to build and collapse. As the field collapses the flux lines cut across the conductors in the secondary coil. This causes a voltage potential that induces current to flow when a load is placed across the secondary coil.

The exact amount of voltage in the secondary coil is determined by the ratio of the number of turns in the primary coil to the number of turns in the secondary. Because the secondary coil has fewer turns than the primary coil, fewer conductors will be available to be cut by the primary's flux lines, inducing less secondary voltage. What this all means to you as an air conditioning technician is that if you put the wrong voltage in the transformer primary you will not get 24 volts out of the secondary.

An easy way to show this is to connect your transformer primary to a variable voltage supply. As you vary the primary volts, the secondary will change in proportion.

ACTIVITY

Have your instructor locate a variable A.C. power supply for you. Connect your primary to the power supply terminals and make the following adjustments in primary voltage. Record the secondary voltage for each change in primary voltage.

Set primary at	Record Secondary volts
10 volts	_____
25 volts	_____
50 volts	_____
110 volts	_____

Now adjust the primary so that there is exactly 24 volts at the secondary. Measure and record the primary voltage _____. In the space below report what you have learned about transformers. Ask the instructor to check your results.

_____

_____

_____

_____

_____

_____

CALCULATING VA RATING

Another important characteristic of a transformer is its *volt ampere rating*. The volt ampere rating is the maximum amount of voltage and amperage that will be available at the transformer secondary. The VA rating is found by multiplying the secondary voltage by the secondary current. Remember, the secondary voltage will stay at 24 volts. The current will change as loads are connected to the secondary. These loads will generally be relay coils. As more coils are added more current is used. Thus, the maximum current must be calculated and measured.

It is important that the VA rating never be exceeded, as this will cause the transformer to burn out immediately.

Two checks must be made to ensure that the transformer's VA rating is not exceeded. The following procedure will lead you through these checks. Check off the steps as you complete them.

1. Locate the VA rating on the data plate of the transformer and record: _____ volt amperes.

2. To determine the maximum current your transformer secondary can yield, complete the following calculation:

    Divide the VA rating by 24 volts.

    VA rating _____ ÷ 24 volts = maximum current _____

    This calculation could also be written

    $24v \sqrt{\text{VA Rating}}$ (The answer is the maximum current allowed on the transformer secondary.)

Now indicate the largest current your transformer can allow _____.

TROUBLESHOOTING THE CONTROL TRANSFORMER

When a malfunction occurs, you may be called upon to correct it. When low voltage is suspected of causing the problem, a complete check of the control transformer and the circuitry leading to its primary winding must be made.

The steps required to isolate a problem in the transformer and its circuitry are listed below. Follow them any time you must troubleshoot the low voltage system.

Ask your instructor to insert a bad component or introduce a problem into the control transformer circuit on your trainer. Then check off each step in the troubleshooting list as you complete it.

Indicate any sub-step that you take by placing an X by its letter.

1. Try to make the unit run by turning on the disconnect, and check the low voltage system for proper operation. (Since there are no loads or relays in the circuit at this time, the presence of 24 volts at terminals R and C means the circuit is operating correctly. The absence of 24 volts indicates a problem.)

2. Record the exact voltage at terminals R and C _____
    This voltage can safely be as high as 28 volts, but a minimum of 24 volts is required.

    a. If the minimum of 24 volts is present you have completed the check of the control voltage transformer and circuit.

    b. If the voltage is below 24 volts or no voltage is present, continue to step 3.

3. Check the primary voltage and record _____
    (Be sure the primary voltage matches the transformer's requirement listed on the transformer data plate.)

    a. If proper voltage is present, continue to step 4. If not, continue to 3b or 3c.

    b. If voltage at the primary is zero, test the voltage source and disconnect switch as described in Unit 2. (When you have returned power to the primary, continue to step 4.)

    c. If primary voltage is present but does not match the voltage required on the data plate, reconnect the transformer primary as outlined in this unit. Continue to step 4 when reconnected.

4. Check terminals R and C again. If you have made changes in step 3, the R and C terminal should now have a minimum of 24 volts.

    a. If no voltage is present at R and C, continue to step 4.

    b. If some voltage is present but it is less than 24 volts, recheck the data plate requirement and the actual primary voltage again. If these voltages match exactly, you have a bad transformer. Remove the bad transformer and replace it with a new one. Then repeat these checks starting with step 1. (Your instructor will provide a new transformer on request.)

5. If step 4 showed no voltage at R and C when correct voltage was present at the transformer primary, you should suspect that one of the transformer coils (windings) has developed an open. Use your ohmmeter to test the coils for continuity as explained earlier in this unit. Indicate which coil is bad _____.

    a. Remove the bad transformer and replace it with a new one. Now repeat this checklist starting with step 1.

DIAGRAMMING THE CONTROL TRANSFORMER

The control transformer, also known as the low voltage circuit, will be diagrammed in both wiring diagram form and ladder diagram form. The wiring diagram will show the exact location of the system's components. This diagram will also show exact routing of all wires.

The circuits in a ladder diagram look like the rungs on a ladder (see figure 4-5)—hence the name. The ladder diagram shows all components in *sequence of operation*. In other words, it shows what happens first in the circuit, what happens next, and so on.

In figure 4-6 the components are positioned the way they are on your trainer. Complete the wiring diagram by drawing in the appropriate wires. Always identify the primary and secondary voltage used for your transformer in the diagram. Have your instructor check your work _____

Figure 4–5

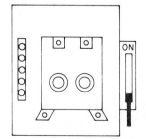

Figure 4–6   Wiring diagram symbols

Complete the ladder diagram when you are finished with the wiring diagram. Always use the correct electrical symbols in a ladder diagram. Figure 4-7 shows the symbols you will need for this diagram.

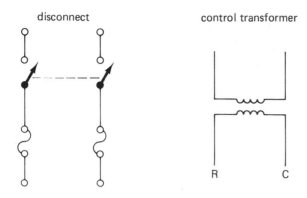

Figure 4-7   Ladder diagram symbols

Complete the ladder diagram so that it shows the operational sequence. Since the disconnect must be closed to power the transformer, draw it first. Then add the transformer below it. Have the instructor check your finished diagram. Instructor's OK _____

This completes Unit 4. When you feel you can fulfill the objectives for this unit, call your instructor for the end-of-unit quiz.

5 Relays and Thermostats

Safety for Unit 5

At times the circuits and disconnects you will be working with will be powered with up to 230 volts. You must be aware of and take appropriate safety precautions relative to electrical shock hazards and wear safety glasses while working.

Tools Required

VOM
Thermostat screwdriver (⅛ in. flat blade)

Objectives for Unit 5

At the conclusion of this lesson the student should be able:

To draw accurately the symbol and identify the terminals of a cooling thermostat.

To install and test a cooling thermostat to the instructor's satisfaction.

To draw the electrical symbol of a relay to the instructor's satisfaction.

To install a relay with the aid of a diagram and control its coil with a cooling thermostat to the instructor's satisfaction.

To draw accurately a wiring diagram of the circuit in objective 4 by tracing the wires in a completed circuit.

To draw a ladder diagram from the wiring diagram in objective 5. This diagram should have the proper symbols and be drawn to the instructor's satisfaction.

THERMOSTAT

Every air conditioning system needs a control device to turn the compressor on when the room gets too warm and off when the room gets too cold. This control device is called a thermostat (see figure 5-1).

The electrical symbol for the thermostat is also shown in figure 5-1. Notice that the symbol combines a switch symbol with the *temperature operator*. The cooling thermostat symbol shows the switch opens down. This means that the switch will close on temperature rise and open on temperature fall.

In this unit, only installation and troubleshooting of a cooling thermostat will be discussed. Complete thermostat operational theory and internal circuitry will be discussed in future units.

INSTALLATION OF THE ROOM THERMOSTAT

The cooling thermostat is a switch. The electrical terminals for this switch are identified as R and Y. The R stands for red, which is the color of the wire that brings power into the thermostat. The Y stands for yellow, the color of the

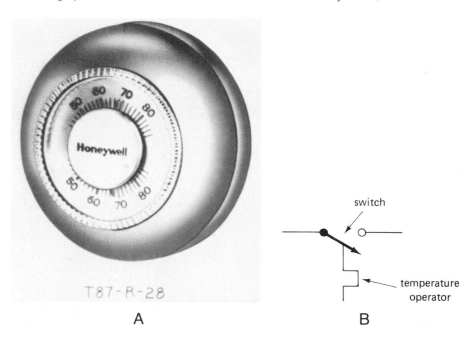

A   B

Figure 5-1   (a) Thermostat (courtesy of *Honeywell Inc.*); (b) symbolic representation wire that takes power from the thermostat to the cooling relay. The cooling relay will be discussed later in this unit. The cooling thermostat will usually be mounted on an inside wall in the conditioned space. A thermostat cable connects the thermostat to the air conditioning unit.

The cable will have two or more conductors as needed by the system. These conductors are connected to a terminal board inside the air conditioning unit. A similar terminal board has been provided on your trainer (see figure 5-2).

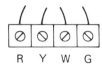

Figure 5-2   Terminal board

The terminals are marked R and Y just like in the air conditioning equipment. There are other terminals marked on the terminal board too (G, W). These will not be used during this unit, but will be explained later.

The installation procedure is outlined below. Follow these steps to install and test your thermostat. Place a check to the left of each procedure as you complete it.

PROCEDURE TO INSTALL AND TEST THE COOLING THERMOSTAT

1. Remove the *cover ring* from the thermostat to expose the mounting screws (see figure 5-3).

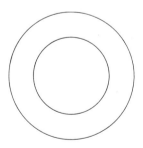

Figure 5-3   Cover Ring

2. Loosen the *three mounting screws* to remove the thermostat from the sub-base (see figure 5-4). Removing the thermostat from the *sub-base* will expose the terminals to which the thermostat wire will be connected (see figure 5-5). Notice that the terminals in the thermostat are marked R, Y, W, and G.

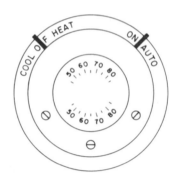

Figure 5-4   Mounting screws

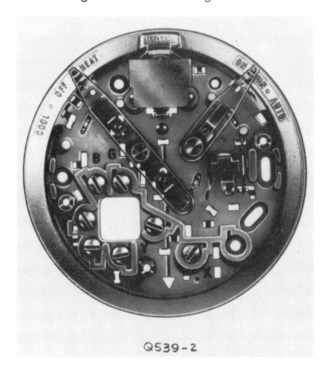

Q539-2

Figure 5-5   Thermostat sub-base (courtesy of *Honeywell Inc.*)

3. Select a piece of 18-4 thermostat wire approximately 20 inches long. Cut the cable insulation back about three inches to expose the four conductors, and strip about ½ inch of insulation off the end of each conductor (see figure 5-6).

4. Bend the bare ends of each conductor in the shape of a question mark (?). This can be accomplished by bending the end of the wire around the shaft of a screwdriver.

5. Loosen the terminals R, Y, W, and G on the sub-base. In this unit only terminals R and Y are being used, but W and G will be connected as far as the terminal board at this time. Place the hooked end of one of the conductors around each terminal in the thermostat. Be aware of color code where possible, connecting the red wire to R terminal, white wire

Figure 5–6   Thermostat cable

to W, green to G, and yellow to Y. Remember to place the hooked end around the terminal in the clockwise direction and tighten the screws to hold the wires.

6.   Check the sub-base to make sure it is level.

7.   Make certain that the bare parts of the wires are not touching each other. Place the thermostat back on the sub-base and tighten the three screws. Two notes of caution:

   a.   Be extra careful not to strip the threads on these screws.

   b.   Be sure the screws are securing the thermostat. These screws make the electrical connection between the thermostat and the sub-base. If they are loose or missing the thermostat will not operate correctly.

8.   Replace the thermostat cover.

9.   Feed the other end of the cable to the terminal board. Strip the insulation and prepare the wires as you did in step 3.

10.   Place the colored wires around the screws in the terminal board. Follow the same color code you used in the sub-base: R for red, Y for yellow, G for green, and W for white.

11.   Use the same procedures as 4 and 5 above to tighten the wires under the screws.

12.   Use an ohmmeter to test R to Y. Set the meter on R X 1 and zero. Turn the dial to the lowest setting. The ohmmeter should show zero resistance, indicating that the switch between R and Y is made. If the reading is infinity (∞) recheck steps 1 to 10.

You have now completed the installation of the thermostat. Since the screws on the thermostat sub-base are fragile, leave the thermostat wired to the terminal board for the rest of the units.

RELAYS

In most air conditioning systems the thermostat controls a relay that switches power on and off to the compressor and fan motor. The switch in the thermostat is too small to control the large current of the compressor directly. For this a *relay* is used. A relay is a magnetically controlled switch.

It is important for the air conditioning technician to understand that the relay has two distinct and separate parts: the coil (see figure 5-7), which becomes a magnet when energized with current, and the contacts (see figure 5-8), which actually comprise a switch. In operation, the coil becomes an electromagnet first, then the contacts (switch) pull closed allowing current to flow to the load. The load is usually the compressor or fan motor.

The circuit that has the coil is called the *control circuit*. The circuit that has the contacts is called the *load circuit*. The coil and contacts can be powered with different voltages.

The electrical symbols for the coil and contacts are shown along with their diagrams in figures 5-7 and 5-8.

To illustrate the fact that the coil and control circuit operate first and independently of the contacts, the coil circuit will be operated by itself during this unit.

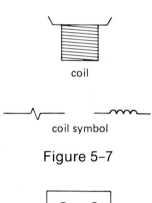

coil

coil symbol

Figure 5-7

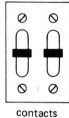

contacts

normally open
contacts

Figure 5-8

The reason air conditioning equipment manufacturers use relays is that a relay coil needs less than 1 amp to become a magnet and pull the contacts open or closed. The contacts can control the very large currents needed to power the compressor and fan.

Typical coil voltages are 24, 120, 208, and 230 volts. This means that the coil can be made to match the control circuit voltage. The control circuit switches, such as pressure switches, can be very small, since only the small current needed to make the coil a magnet will pass through their contacts.

INSTALLATION OF A RELAY

As an air conditioning technician you will be required to install and/or test relays in equipment. The following procedure will be used for both installation and testing of all relays. Use the procedure to install and test the relay on your trainer. Check off the steps as you complete them. For this installation a relay with a 24 volt coil will be used. (Be sure power is off when making the connections.)

1. Locate the relay which has a 24 volt coil on your trainer. The coil voltage is marked right on the coil.

2. Connect a control voltage transformer to the proper voltage. (Refer back to Unit 4 if you need help.)

3. Check the diagram for complete circuits. Use the diagrams to complete step 4 (see the ladder and wiring diagrams of figure 5-9).

4. Connect the low-voltage circuit wires as per diagram, starting at terminal R of the transformer.

   a. Connect a red wire from terminal R on the transformer to R of the thermostat terminal board. (Use proper connecting devices.)

   b. Wire from Y of the thermostat terminal board to either coil terminal (be sure the coil is 24 volts).

   c. Connect a wire from the remaining coil terminal of the relay to the C terminal of the transformer.

5. Power the circuit by turning the disconnect switch on.

   a. Switch the thermostat to cooling and adjust the temperature setting to the lowest possible number. If the coil pulls the contacts in, continue to step 6; if it does not, continue to step 5b.

   b. If the contacts do not pull in, move the thermostat setting to the highest number, and then return to the lowest setting and leave it. If contacts still do not pull in, use a voltmeter and check for 24 volts A.C. across the coil terminals (see figure 5-10 for proper meter probe placement).

   c. If voltage is 24 volts, the circuit is wired properly and the transformer and thermostat are working correctly. *The problem is with the relay coil.* Continue to step 5e.

   d. If voltage is less than 24 volts return to step 2 and recheck. (You are looking for proper circuit connections and proper transformer voltage at R and C.)

   e. Recheck the coil rating. It should be 24 volts A.C. If the rating is not 24 volts, repeat the procedure from step 1. If the rating is 24 volts, continue to step 5f.

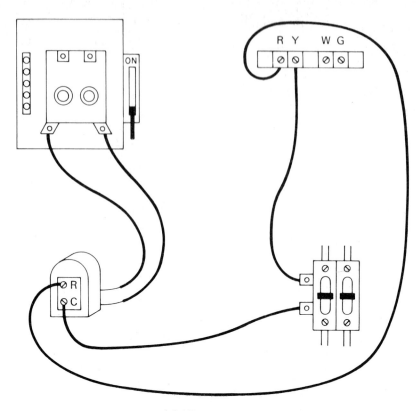

(a) Wiring diagram

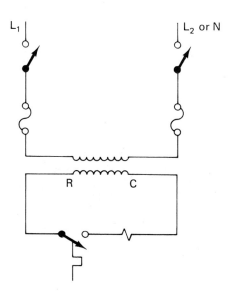

(b) Ladder diagram figure

Figure 5-9   (a) Wiring diagram; (b) ladder diagram

*Relays and
Thermostats*

48

f.   Check the coil for continuity. Use an ohmmeter and set the range to X100 K. If reading is infinity, the coil has developed an open and must be replaced. If reading shows the coil has continuity (some resistance), call the instructor to watch you repeat steps 1 through 5.

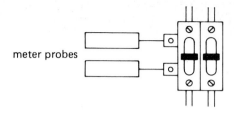

meter probes

Figure 5-10   Meter probes across the relay coil

The previous test demonstrates that the coil can operate even when no power is connected to the contacts. The coil simply closes the contacts.

Before you remove the wires from your trainer, complete the wiring diagram below (see figure 5-11). To complete this diagram follow your wires from point to point on the electrical components and draw these wires as lines on your diagram. Indicate the wire colors as you proceed. Have your instructor approve your diagram before you continue. Instructor's approval _____

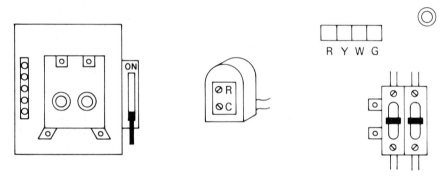

R Y W G

Figure 5-11

Once you have completed the wiring diagram, use it to draw a ladder diagram of your circuit. Review the ladder diagram in the previous unit if necessary. Have your instructor approve your diagram _____.

You have now completed Unit 5. When you are ready to demonstrate that you have fulfilled the objectives of the unit, call your instructor for the end-of-unit quiz.

Installation of a Relay

6 Basic Hermetic Motor Theory

Tools Required

VOM
Clamp-on ammeter (100 amp range)
#2 flat tip screwdriver

Objectives for Unit 6

At the conclusion of this lesson the student should be able:

To explain to the instructor's satisfaction the operating theory for an A.C. single-phase induction motor.

To sketch accurately the relative locations of the start winding and the run winding in the stator.

To explain orally or in writing the difference between the start and run windings of a single-phase induction motor.

To identify with 100% accuracy the terminals of a single-phase hermetic compressor with an ohmmeter.

To draw the electrical diagram of a single-phase induction motor (start and run windings) to the instructor's satisfaction.

THE HERMETIC COMPRESSOR

In most air conditioning and refrigeration systems the compressor motor and pump are sealed in a container (see figure 6-1). When a compressor is sealed in this manner it is called a *hermetic compressor*. The motor is sealed so that the system's refrigerant can be used to cool the motor windings without being lost to the atmosphere.

As an air conditioning and refrigeration technician, you will not open the hermetic compressor for maintenance. If the motor fails, the hermetic compressor is replaced as a unit. Since the hermetic compressor on your trainer cannot be opened, you will disassemble an open motor to inspect its basic parts (see figure 6-2). The open motor parts will be very similar to those in the hermetic compressor.

INSPECTION OF THE SINGLE PHASE INDUCTION MOTOR

The A.C. single-phase induction motor has three main parts: the *rotor*, which is the rotating part (see figure 6-3), the *stator*, which is the stationary part (see figure 6-4); and the *end plates* that support the rotor (see figure 6-5).

Figure 6-1   Hermetic compressor (courtesy of *Copeland Corp.*)

Figure 6-2   Open type motor (courtesy of *General Electric*)

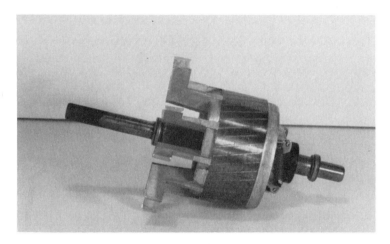

Figure 6-3   Rotor (courtesy of *General Electric*)

Figure 6-4   Stator (courtesy of *General Electric*)

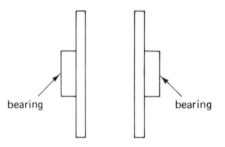

Figure 6-5   End plates

To get a closer look at these parts, disassemble the single-phase induction motor using the following procedures.

Check off the steps as you complete them.

1. Locate the single-phase induction motor pictured in figure 6-2.

2. Loosen the mounting screws and remove the motor from the trainer. Place the motor on a clean work table for disassembly.

3. Before the motor is disassembled by removing the end plates, it is important to mark the end plates and stator so that the end plates will be reassembled in the exact original position. The motor is marked by making a scratch mark about ¾ inch long as shown in figure 6-6. The mark should start on the end plate and end on the stator. Use the blade of a screwdriver to make the scratch. Check the motor for previous marks. (If marks are already present, do not mark the motor again.)

4. Remove the four nuts that hold the end plates onto the stator. Then remove the four bolts. Be careful where you keep these parts, as you will need them for reassembly.

5. Insert a screwdriver between the end plate and the stator. Pry gently to remove the end plate.

6. Carefully pull the rotor out of the stator and set the rotor and end plates aside.

7. Look closely at the stator. Notice that it is made of coils of copper wire. The copper wire has a clear insulation on it. Never mark or crack this insulation. When a motor fails, it is usually because the insulation is worn from being hot due to too much current. This may cause a short circuit between the wires in the coils.

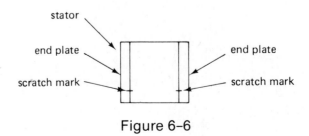

Figure 6-6

8. Locate two different-size windings in the stator. The coil with the heaviest wire is called the *main* or *run winding*. The coil with the lightest wire is called the *auxiliary* or *start winding*.
   Sketch the location of the start and run winding in your stator here (notice they are offset):

9. Inspect the rotor. Notice that it does not contain any wire. In an A.C. induction motor, the rotor is made by pressing laminated steel onto an iron core. The iron core looks much like an exerciser in a squirrel's cage. For that reason the rotor is sometimes called the "squirrel cage" and the induction motor is sometimes called a "squirrel cage motor."

10. Inspect the end plates. This is the only part that will be somewhat different in the hermetic motor. Locate the bearings in the end plates. Their main job is to hold the rotor in position and allow the rotor shaft to rotate with a minimum of friction.

11. Replace the rotor into the stator so that the shaft comes out the end opposite the place where the wires come through the stator.

12. Put the end plates back on the rotor and replace the four bolts that hold the end plates together.

13. Replace and tighten the nuts on the bolts. Be sure the rotor can spin freely. If the rotor can not spin freely, call your instructor. You have completed the disassembly and inspection of your motor. Return the motor to the trainer and tighten the mounting screws.

OPERATIONAL THEORY OF SPLIT-PHASE INDUCTION MOTORS

As an air conditioning and refrigeration technician, you must understand motor theory in order to troubleshoot the single-phase induction motor. The main job of a motor is to turn the shaft. The rotor is mounted on the motor shaft, so the rotor must spin to turn the shaft. The basic operation of the induction motor is very similar to that of the simple D.C. motor you studied prior to starting this course. The rotor and stator are magnetized so that the attraction and repulsion will cause the rotor to spin. The main difference between the A.C. and the D.C. motor is that the rotor in the A.C. motor gets its current for magnetization by induction from the stator rather than from the brushes as in the D.C. motor. To do this the rotor is made by laminating layers of steel that are pressed on the "squirrel cage."

The induced current is similar to that in a transformer. Think of the stator winding as the primary coil and the rotor as the secondary coil. The magnetic field builds in the stator and as it collapses (due to the alternating current) it induces a current into the rotor causing a magnetic field to build there. A second magnetic field in the stator will rotate if the rotor is given a spin. Once this field begins to rotate in the stator, the rotor will "chase" or follow the movement of this rotating field. This causes the rotor to spin.

You can demonstrate this theory on your single-phase induction motor using the following procedure. Check off the steps as they are completed.

1. Locate the single-phase induction motor on your trainer and record the voltage listed on the motor's data plate _____.

2. The run winding of your motor is connected to the black and white wire that comes out the side of the motor. Find the black and white wire on your motor and prepare their ends by stripping back about ½ in. of insulation.

3. Put a wire nut on the ends of any other wires coming from the motor (your instructor will supply these nuts.)

4. Make sure the single phase disconnect is turned off.

5. If the voltage rating for your motor is 115 volts, connect the white wire to the load side terminal of L_1. Connect the black wire to the neutral bar (see figure 6-7). Do not turn the switch on yet!

    a. If the voltage is 208 or 230 volts, connect the black wire to the load side terminal of L_1 and the white wire to the load side terminal of L_2. Do not turn the switch on yet!

6. Read this complete step before applying power to your motor. You are trying to show that the rotor or shaft of the induction motor will rotate and follow the rotating field in the run winding. The rotor must be started to get the magnetic field to rotate. To do this you must apply power and give the shaft a spin in either direction. *Be aware of a safety hazard. Watch your hands and loose clothing when the shaft begins to spin so they don't get caught in the shaft.* If you spin the shaft and the motor does not begin to run, turn the disconnect switch off immediately and call your instructor.

    Now that you understand these points, apply power by turning on the disconnect and giving the shaft a spin. If the motor does not continue to run, turn the power off immediately.

    Now you can see that once the rotor starts to spin it will continue as long

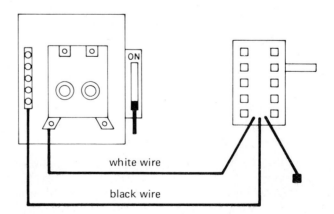

white wire

black wire

Figure 6-7   Wiring diagram for 115 volt motor

as there is current in the run winding. The reason you must give the rotor (shaft) a spin is to start the magnetic field in the stator rotating.

Giving the shaft a spin every time you wanted the motor to start would not be practical. However, a start winding that is placed in the stator 90° from the run winding will start the rotor moving (see figure 6-8). The start winding is made of very fine wire with many turns to make the magnetic field very strong. This gives the rotor as much torque as possible on start-up. Remember from the basic D.C. course that torque is the rotating force the shaft will have.

The start winding is made of very fine wire and will draw a very large current on start-up. Since the rotor has not started to turn, the amperage is called *locked rotor amperage* or *LRA*. This amperage would burn up the start winding if the winding were left in the circuit very long. For this reason the start winding must be disconnected when the rotor reaches approximately 75% full speed (around two or three seconds). To demonstrate this theory use the following procedure. Check off the steps as you complete them.

Make sure the disconnect is turned off before you proceed to these steps.

1. Locate the brown wire on your motor and remove the wire nut on its end. The brown wire is the start winding of your motor.

2. Connect the brown wire to one side of a normally open push-button switch (Your instructor will provide the switch.)

3. Connect a wire to the same disconnect terminal as the white wire of the motor. Connect the other end of that wire to the pushbutton switch (see figure 6-9).

4. *Read this step completely before you turn on the power.* This time, instead of giving the shaft a spin, you will demonstrate that the start winding will cause the rotor to begin spinning. To do this you will energize the start winding for one to two seconds by pushing closed the pushbutton switch when the disconnect is turned on. When you let up on the pushbutton it will automatically open (disconnect) the start winding. If the motor does not start, be sure to let up on the switch and turn the disconnect off immediately. When you understand this step, turn the disconnect on and close the pushbutton for just a second. Remember if the motor fails to start, turn the power off at the disconnect immediately and call your instructor.

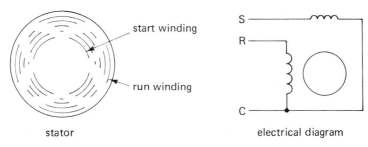

stator   electrical diagram

Figure 6-8   (a) Stator; (b) electrical diagram

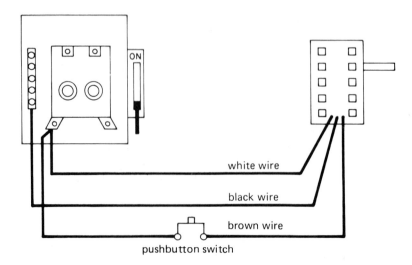

white wire

black wire

brown wire

pushbutton switch

Figure 6-9

5. If the motor starts and continues to run, turn it off and repeat step 4 several times, making certain that the motor starts each time. Turn the disconnect off and continue to step 6.

6. It is important to measure the locked rotor amperage when the motor is started. A clamp-on ammeter will be used to measure this amperage (see figure 6-10).

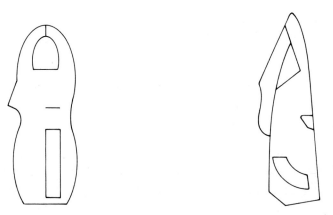

Figure 6-10   Clamp-on ammeters

To measure the LRA open the jaws of the clamp-on ammeter and slip them around the black wire of your motor (see figure 6-11). Roll the dial around to set the meter range to 100 amps.

7. Now repeat step 4 with the clamp-on ammeter in place. You should notice that the current goes very high for a second and returns to about 5 to 10 amps. This high current is the locked rotor amperage.

Record the LRA _____

The point where the current returns to the 5 to 10 amps is called the full load amperage (FLA).

Record the FLA _____

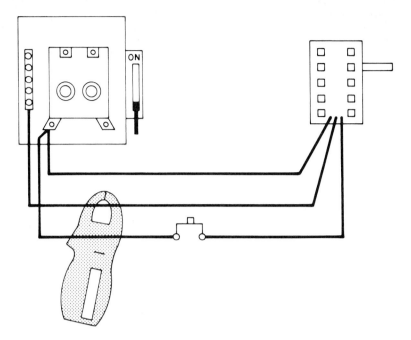

Figure 6-11   Ammeter in place

8. After you have measured and recorded the full load amperage, turn the power off and disconnect all wires. Continue to the next part of this unit.

IDENTIFYING THE TWO WINDINGS IN THE STATOR

As you noted in your inspection, the stator has two different-sized windings. The winding made of the largest wire is called the *main* or *run winding*. The wire is large so it can handle the heat and current that occurs when the motor is running.

The other winding made of the smaller wire is called the *auxiliary*, or *start winding*. Since its wire is small, the start winding can only be in the circuit for a few seconds, during the starting operation. Because of this the single-phase induction motor will always have a means of disconnecting the start winding.

In the hermetic motor one end of the run winding and start winding is soldered together. The point where the two ends are soldered becomes a new terminal, called *common* (see figure 6-12).

Figure 6-12

Identifying the Two Windings in the Stator

59

On some compressors, a device called an external overload (see figure 6-13) will limit the maximum current to the run winding and the start winding. The external overload is generally connected to the common (C) terminal on the compressor. If the full load amperage gets too high the overload will open, turning off the compressor. After the overload cools off it will close again and the compressor will restart. The overload should prevent the compressor from burning out due to excessive current. The overload can be tested for continuity (like a switch) with an ohmmeter. If the overload is open, allow approximately 10 minutes for it to cool down and close.

Ex OL

Figure 6–13   External overload

Since you can not open and examine the hermetic motor, it is important that you be able to identify the run, start, and common terminals in the hermetic compressor with an ohmmeter. Locate the hermetic compressor on your trainer and use the following procedure to identify its terminals.

1. Locate the three terminals on your hermetic compressor.

2. Make a sketch of the three terminals, carefully noting their correct position in the sketch.

3. Set the ohmmeter on R × 1 and zero the meter.

4. Using two terminals at a time, test and record the resistance between the compressor terminals. Be careful to record the measurements onto your sketch in the exact location that you measured them. The amount of resistance will vary, but usually it will be less than 20 ohms. Be sure to make these measurements as accurately as possible.

5. An example will be used here to show you how to identify the terminals after you have added resistance readings to your sketch.

```
              1
            / \
          5 Ω   8 Ω
          /       \
        3 —13 Ω— 2
```

a. To identify the terminals in the example, find the *highest* reading. It is 13 ohms.
    The two terminals that gave us 13 ohms are terminals 3 and 2. The terminal not used (1) is common; label it C in the diagram.

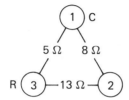

b. Now, look for the *lowest* reading. In the example, this is 5 ohms, occurring between terminals 3 and 1. Mark terminal 3 as R for run. We can determine this because we know the run winding is made of larger wire and fewer turns than the start winding and will have the least resistance.

Add this to the diagram

c. This leaves terminal 2 as start (S). To double check our answer we should refer back to this diagram.

Add the resistance from our example.

We can see that from R to C the resistance 5 ohms is the smallest reading and therefore the run winding.
S to C: 8 ohms, larger than the run winding.
R to S: 13 ohms, the sum of 5 and 8, and always the largest reading.

6. Now use these steps to identify the terminals in your sketch.

    a. Find the two terminals that yield the largest resistance reading. On the remaining terminal mark C for common.

    b. Find the two terminals that yield the lowest reading. One terminal should be C (common), the other R for (run).

    c. The remaining terminal is S. Recheck to be sure that R to C is the lowest reading and R to S is the highest reading.

Use this method any time you must identify the compressor terminals.

You will use a similar method when you are troubleshooting the compressor itself. Usually the problem will be an opening in one of the windings. Troubleshooting the compressor will be covered in the next unit.

You now have completed Unit 6. When you are ready to take the end-of-unit quiz, call your instructor.

7 The Split Phase Compressor

Tools Required

Voltmeter (0-250 volt scale)
Screwdriver (flat blade ¼ in.)

Objectives for Unit 7

At the conclusion of this lesson the student should be able:

To explain either orally or in writing to the instructor's satisfaction how the split phase compressor starts.

To identify correctly the terminals M, S, and 1, given a current relay.

To draw the diagram of a current relay correctly.

To explain either orally or in writing the operation of a current relay.

To test correctly the coil and contacts of a current relay using an ohmmeter.

To wire correctly the current relay to the split phase compressor and test run, given a diagram of a split phase compressor and a current relay.

To wire and test run the circuit, given a diagram, of a cooling thermostat controlling a split phase compressor.

SPLIT PHASE COMPRESSORS

Hermetic compressors are used in most air conditioning and refrigeration units. The hermetic motor can be wired in any one of four ways to produce the required starting torque. These types of wiring include (1) *split phase* for low torque, (2) *permanent split capacitor* for medium torque, (3) *capacitor start-induction run*, and (4) *capacitor start-capacitor run* for high torque. This unit and the next three units cover procedures to show the theory of operation, installation, troubleshooting, and repair for each of these four types of compressors.

This unit will cover the hermetic compressor wired as a split phase motor. It will be referred to as the split phase compressor.

The name split phase is used because the motor uses the phase shift that occurs between the run and start winding currents to start the motor. This means that the current in the start winding is out of phase (split) with the current in the run winding when the motor is started (see figure 7-1). This is achieved by having the two windings made of different size wire and set in the stator at 90° to each other. This split in the current phases causes the magnetic field in the stator to rotate. The rotating field causes the rotor to begin to move. The amount of force the shaft can produce during starting is called *starting torque.*

Air conditioning and refrigeration manufacturers use the split phase compressor in small units that need low starting torque, for example window air conditioners, some household (domestic) refrigerators, and dehumidifiers.

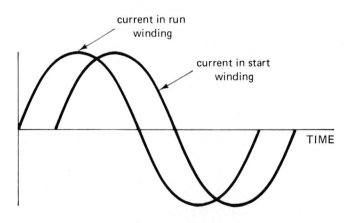

Figure 7-1   Phase shift

THE CURRENT RELAY

The start winding can only stay in the circuit for a few seconds during starting since it will draw a large current. A switch is used to take the start winding out of the circuit when the motor reaches 75% of full RPM. This switch is called a *current relay* (see figure 7-2). Two types are shown.

The current relay has a coil and one set of normally open contacts (see figure 7-3). The coil is made with a few turns of large wire to withstand the large run winding current during starting. Notice in the diagram that the contacts are connected to one side of the coil.

The split phase compressor depends on the correct operation of the current relay for starting. For this reason it is important that you know how to test the current relay.

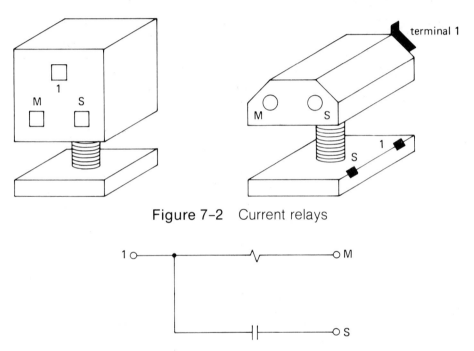

Figure 7-2   Current relays

Figure 7-3   Symbol for current relay

TESTING THE CURRENT RELAY

Use the following procedure to test your current relay. Check off the steps as you complete them.

1. Locate the current relay on your trainer. Notice that the terminals are marked S, M, and 1

2. The relay coil should be located between terminals 1 and M (refer to figure 7-3). Use your ohmmeter to test the relay coil. Set the meter R × 1. Put one probe on M and one probe on terminal 1.

    Record the measurement _____.

   a. If the meter shows the coil has some resistance, the coil is good. Continue to step 3.

   b. If the meter shows infinite (∞) resistance the coil is open and the relay must be replaced. Ask the instructor to give you a new relay and repeat step 2.

3. The contacts are between terminals 1 and S. Use your ohmmeter to test the contacts. Place one meter probe on S and one probe on terminal 1. (Note: this relay is position-sensitive. The arrow on the side of the relay will show position. For this relay the arrow should point up when the relay is in the proper position.)

   a. With the arrow pointing up, the contacts should be open. Set the meter to the highest resistance scale. Measure and record the resistance between terminals 1 and S on the relay _____. If the reading is infinity (∞) continue to step b. If the reading is not infinity the contacts may be welded or sticking, and the relay needs to be replaced. Ask your instructor for a new relay and repeat step 2.

   b. Turn the relay over so the arrow points down and repeat the resistance measurement. This time the contacts should be closed and show less than 1 ohm resistance. The resistance between terminals 1 and S is _____. If the resistance in more than 1 ohm, replace the current relay and repeat the test procedure. Remember, a good current relay will show low resistance in its coil, zero resistance with its contacts closed, and infinite resistance when its contacts are open.

STARTING THE SPLIT PHASE COMPRESSOR

The current relay uses the large surge of starting current in the run winding to magnetize its relay coil. The starting current, *locked rotor amperage* (LRA) increases to a peak of three to five times the full load amperage for about two seconds (see figure 7-4). The magnetism pulls the relay's contacts closed to complete the circuit in the start winding (see figure 7-5). As current flows in the start winding the magnetic field in the stator begins to rotate. The motor's rotor follows this rotating field. As the rotor picks up speed it begins to produce *counter-electromotive force* (CEMF).

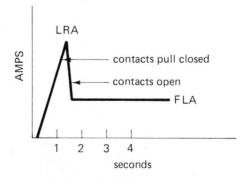

The Split Phase Compressor

Figure 7-4   Starting current diagram

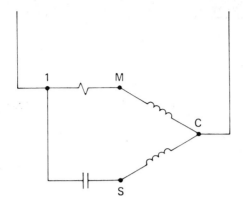

Figure 7-5 Current relay on motor

The CEMF reduces current in the run winding. When the current decreases to near FLA, the relay contacts drop open, stopping current flow to the start winding. The rotor, now started, will continue to turn until power is disconnected.

TEST RUNNING THE SPLIT PHASE COMPRESSOR

Use the following procedure to test run your split phase compressor.

1. Locate the hermetic compressor on your trainer.

2. Remove the terminal cover and set it aside (it will be replaced later).

3. Locate the current relay and install it by pushing it on the run and start terminals of the compressor (see figure 7-6). (If you need help to identify the terminals on the compressor, use the procedure in Unit 6.)

4. Locate the data plate on the compressor and record the required voltage _____.

5. Locate the single phase disconnect and turn off its switch. Use your voltmeter to be sure the power is off.

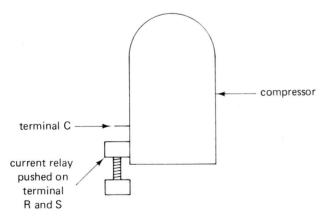

Figure 7-6 Current relay on compressor

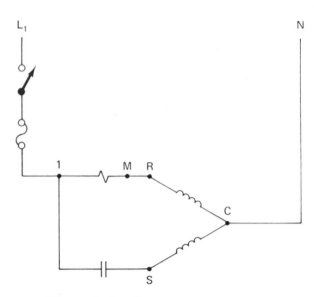

Figure 7-7 Current relay on motor

6. If your compressor needs 110 volts, follow the diagram in figure 7-7 as you connect terminal 1 of the current relay to the L_1 load side terminals in the single phase disconnect (watch the color code for the conductors). Connect the C terminal of the compressor to the neutral bar in the disconnect.

7. If the compressor is 230 volts, use L_1 to terminal 1 on the relay and connect terminal C of the compressor to L_2 in the disconnect.

8. Place the clamp-on ammeter around the conductor on terminal C of the compressor. Set the meter to the 30 amp scale

9. Read this step carefully before turning on the power:

    As you turn the disconnect on, watch the ammeter. The scale should show the LRA current for about two seconds, and then the current should decrease by about two thirds. (For example, the current goes to 15 amps at start and then returns to 5 amps.) If the current does not decrease to the full load amperage level once the compressor starts, turn the disconnect off immediately. If the current returns to the full load level, allow the compressor to run. Now you are ready to complete step 9. Turn on the power in the disconnect.

10. If the compressor fails to start or continues to draw excessively high current, turn the power off immediately and call your instructor to check your wiring and components.

CONTROLLING THE SPLIT PHASE COMPRESSOR

The Split Phase Compressor

In some air conditioning equipment, the split phase compressor will be turned on and off at the appropriate time by a line voltage thermostat (see figure 7-8). The thermostat is in series with the L_1 conductor. The National

Electrical Code specifies that the neutral line can never have a switch in it. For this reason the thermostat will be in the "hot" line that connects the L_1 terminal on the motor's current relay. This operation of this thermostat will be covered in later units. Use the following procedure with the diagram in figure 7-9 and wire your split phase compressor as it would be found in manufacturer's equipment.

Figure 7-8   Electrical symbol of thermostat

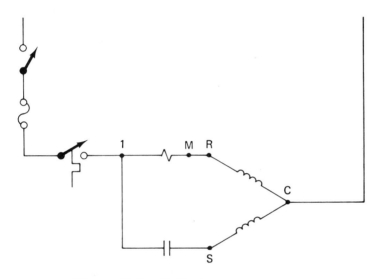

Figure 7-9   Split-phase compressor

1.   Locate the compressor on your trainer and connect the current relay to its terminals as you did in the previous procedure.

2.   Check the compressor's data plate to determine the voltage required to operate your compressor.

Record the voltage required _____

3.   Locate a single-pole single-throw switch on your trainer. It will look like the switch used to turn off the lights in your room. This switch will be used as your thermostat. You will need to turn it off and on by hand. (An actual thermostat would turn off and on by the changing temperature.)

4.   Turn off all power while installing the wiring.

5.   Follow figure 7-9 and make the connections as they are listed.

   a. Connect a wire from L_1 load side terminal of the single phase disconnect to the switch (use either of the two screws).

b. From the other screw connect a wire to terminal 1 on the compressor (remount the switch on its 2 in. × 4 in. electrical box).

c. From the C terminal on the compressor, connect a wire to the neutral bar in the single phase disconnect.

6. Place the clamp-on ammeter on the wire that goes to terminal 1 on the current relay. Set the scale to the 30 amp range.

7. Turn on the disconnect and turn on the switch and watch the current to be sure the motor starts and the current relay opens the start winding. Remember that if current stays high and does not return to 6-10 amps, *immediately shut off* the power and call your instructor.
If the current returns to a normal FLA level after start, allow compressor to run for a few minutes and record the amperage _____.

8. You have completed this test. Turn the power off and leave the circuit in operating condition. Ask your instructor to put a problem in your circuit for the next section on troubleshooting.

TROUBLESHOOTING THE SPLIT PHASE COMPRESSOR

When the split phase compressor fails to start, the air conditioning and refrigeration technician must be able to troubleshoot the circuit and components quickly and accurately. The following procedure lists the steps in troubleshooting the split phase compressor. These steps will be broken into two parts.

Ask your instructor to put a problem in your circuit. (Possible problems: bad wiring, bad current relay, bad windings in compressor, no voltage.) Check off the following steps as you complete them. Now refer to figure 7-9 to complete this test.

1. Put a clamp-on ammeter on L_1 and try to make the system run. Turn the disconnect switch and thermostat switch on. Check the current draw as you try to start the system. If the current becomes excessive and the motor will not start, turn the power off *immediately* and go to steps 2 and 3.

a. If the compressor runs and the full load current is within limits, you have completed the test (ask for another problem).

2. If the compressor tries to run (makes a noise) but does not start, skip to step 4. (Clue: the current will be high.)

3. If the compressor does not try to run (does not make a noise), skip to step 5. (Clue: no current will be present.)

4. Since your compressor is "trying" to start, it means that the circuit and thermostat are probably all right. The problem is then in the current relay, or in the compressor motor or connections. Use the procedure listed in this unit to test the current relay, and record the results of your test.

a. The current relay is good. Go to step c.

b. The current relay is defective. Install a new current relay and test run the compressor.

c. Use the procedure listed in Unit 6 to test the start and run windings of the compressor and record your results:

Compressor is bad _____

Compressor is OK; go to step d _____

d. Inspect the terminal of the compressor and current relay for dirty or loose connections. Try to make the compressor run again. If the compressor still fails to start, call your instructor to help you repeat the above steps.

5. If the compressor does not make a noise and current draw is zero, the problem is in the circuit and/or components supplying the power to the current relay and terminal C on the compressor. The other possibility is an open in terminal C inside the compressor.

6. Use a voltmeter for this test. Place one voltmeter probe on the L_2 or N of your circuit and leave it (L_2 for 208 volt circuit, N for 115 volt circuit).

7. Move the other probe around the circuit to the test points listed in 7a, b, c, d, e, and record the voltage you measure. When your meter measures proper voltage, it means that part of the circuit is OK. Continue checking until you find the point where voltage is zero. The problem will be between the point where voltage is zero and the last point where full voltage was measured. Remember, one probe must stay on L_2 or N for this part of the test to work. (If full voltage is found all the way to terminal R on the compressor, then the left side of the circuit is OK. Reverse the probes and check the right side of the circuit.)

8. Test points for left side of circuit:

a. Line side of disconnect (if no voltage, problem is in power supply).

b. Load side of disconnect (if no voltage, fuse is bad).

c. Line side of thermostat switch (if no voltage, wire between disconnect and thermostat is bad).

d. Load side of thermostat switch (if no voltage, thermostat is bad).

e. Terminal 1 on current relay (if no voltage, wire between thermostat and current relay is bad).

If full voltage is measured at terminal 1 on the current relay, the left side of the circuit is good. Move the probe from L_2 or N to L_1 and leave it. Use the same method to test the right side of the circuit. This means one probe will stay on L_1 and the other probe will be moved to test the following:

f. Load side of L_2 (if no voltage at L_2, fuse is bad).

g. Terminal C on compressor (if no voltage, wire between L_2 in disconnect and terminal C is bad).

If power is present at terminal C on compressor and terminal 1 on the current relay, the compressor is bad. Use the test method listed in Unit 6 to confirm this suspicion.

This completes the troubleshooting test. Turn off the power on the trainer. Disconnect all wires and components and put them on their proper storage location.

You have now completed Unit 7. Call your instructor when you are ready to take the quiz on Unit 7.

8 The Capacitor-Start, Induction-Run Compressor

Safety for Unit 8

At times the circuits and disconnects you will be working with will be powered with up to 230 volts. You must be aware of and take appropriate safety precautions relative to electrical shock hazards and wear safety glasses while working. A start capacitor will be used during the lesson. You should follow instructions provided in this unit to discharge and handle the start capacitor.

Tools Required

Voltmeter (0-250 volt scale)

Screwdriver (flat blade ¼ in.)

Clamp-on ammeter (30 amp capacity)

Objectives for Unit 8

At the conclusion of this lesson the student should be able:

To draw accurately the electrical symbol for a capacitor.

To test a start capacitor for an open or short circuit using an ohmmeter.

To draw accurately a ladder diagram of a capacitor start-induction run motor from a wiring diagram of a CSIR (capacitor start induction run) motor.

To explain to the instructor's satisfaction the theory of operation of a capacitor.

To explain to the instructor's satisfaction the theory of operation of a CSIR compressor.

Given a diagram, to install a start capacitor and current relay on a hermetic compressor and test run.

To diagnose accurately an electrical malfunction in a CSIR compressor's electrical system.

THE CAPACITOR-START, INDUCTION-RUN COMPRESSOR

When hermetic compressors need more torque to get started, a *start capacitor* is added to the split phase compressor in series with the start winding. After a start capacitor is added, the split phase compressor is known as a *capacitor-start, induction-run compressor* (CSIR).

The start capacitor (see figure 8-1) is made of two conducting plates separated by an insulator and placed in a black plastic container. The start capacitor is rated in *microfarads* (uf). Microfarads may also be written as mf and MF. The typical rating of start capacitors is between 75 uf and 600 uf.

When power is applied to a start capacitor, one of the conducting plates fills with electrons and stores a charge equal to the applied voltage. The electrons on the charged plate have a great attraction to the uncharged plate. The electrons cannot flow through the insulation, also called the dielectric, to get to the uncharged plate. Instead, the electrons must flow out the charged plate's terminal and travel through an external circuit to get to the uncharged plate. In the case of the capacitor start induction run compressor, the external circuit is the start winding.

The capacitor can become defective in several ways. The dielectric can become weak and the capacitor lose its ability to store the charge, or the plates can short circuit. You can test the capacitor with an ohmmeter for

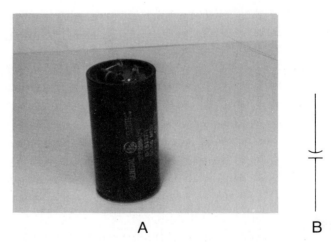

A   B

Figure 8-1   (a) Start capacitor (courtesy of *General Electric*); (b) electrical symbol

each of these problems. Use the following procedures to test the start capacitor on your trainer.

The following procedures can be used to test a capacitor with an ohmmeter. Check off the steps as you complete them.

1. Locate the start capacitor on your trainer. If your capacitor does not have a resistor across its terminals, use a resistor of any value under 1 K to discharge the capacitor. Discharging the capacitor is accomplished by placing the ends of the resistor across each of the terminals on the capacitor to complete the circuit, allowing the voltage to flow from one plate to the other. Failure to discharge the capacitor before touching its terminals may result in the capacitor discharging in the technician's hand. This is not a life threatening voltage, but the reaction of the technician to the spark is usually to jump away, out of control, and a dangerous fall or other accident could occur. If a resistor is soldered across the terminal this step may be omitted.

2. If the capacitor has a resistor soldered across its terminals it must be carefully unsoldered from one terminal to complete this test. At the end of this test the resistor must be carefully resoldered. This resistor, called the *bleed resistor*, will automatically discharge the capacitor when the current relay opens. This will leave the capacitor discharged when the technician must work on the system while the power is turned off.

3. Set your ohmmeter on R × 10 and zero the meter.

4. To test for opens and shorts in the capacitor, touch the meter probes on the two capacitor terminals. The capacitor should charge up from the voltage of the battery in the meter. Since the battery is D.C. voltage you must reverse the probes on the terminals for the capacitor to charge again. This means that if the left terminal was on the capacitor for the first test, it should touch the right terminal for the next test. When the probes touch the capacitor terminal the meter needle should swing toward zero, then slowly fall back to infinity (∞). If the meter needle does not move, change to a different meter scale (R × 1 or R × 1K) and reverse the meter leads when you touch the terminals again.

    a. If the meter needle does not move at all, the capacitor has an open circuit, meaning there is not a conducting path between the plates and the terminal. The capacitor is bad and needs to be replaced. Ask your instructor for another capacitor.

    b. If the meter needle moves to zero but will not swing back to infinity (∞), it means the capacitor has a short between its conducting plates. The capacitor is bad and must be replaced. Ask your instructor for another capacitor.

    c. If the meter needle moves toward zero and then falls back to infinity (∞) the capacitor is good. When the needle moves toward zero it shows the capacitor is charging, and when the needle falls back to infinity it shows the capacitor is discharging.

You have now completed testing your start capacitor for open shorts and charging. Carefully resolder the resistor across the terminals if it had to be removed for this test.

STARTING THE CAPACITOR START, INDUCTION RUN COMPRESSOR

The capacitor start, induction run motor is essentially a split phase compressor with a start capacitor is series with its start winding (see figure 8-2).

The CSIR motor needs the current relay to remove the start winding from the circuit just as in the split phase motor.

During the starting of the CSIR motor, the current relay coil pulls its contacts closed. This allows one plate of the start capacitor to charge to the voltage level of the applied voltage. This will occur during the first half cycle of the applied voltage sine wave. As the sine wave changes to the second half cycle, the voltage flow is reversed, and current flows out of the capacitor back through the run winding and then to the start winding to get to the other plate of the capacitor.

Note that voltage does not pass between the two plates of the capacitor; rather it charges one plate, then discharges that plate through the motor winding to the other plate. This charging and discharging is in time with the changing of the sine wave of the applied A.C. power.

When the voltage is forced to charge the capacitor, then discharge through the motor winding, it causes a larger phase shift between the voltage in the start and run winding than occurred in the split phase motor. This larger phase shift gives the CSIR motor more starting torque so it can start against larger loads.

Since the CSIR compressor has more starting torque than the split phase compressor, manufacturers use the CSIR compressor on equipment that has expansion valves as the metering device, such as commercial refrigeration units and some domestic refrigerators and freezers. Since the expansion valve equalizes the pressure very slowly during the off cycle, the compressor usually has to start against a high head pressure. By having a start capacitor, the CSIR compressor can easily start against the high head pressure.

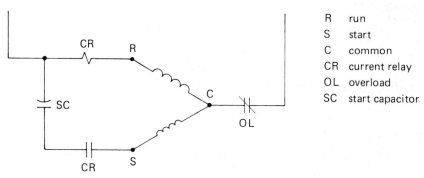

R   run
S   start
C   common
CR   current relay
OL   overload
SC   start capacitor

Figure 8-2   Capacitor-start, induction-run compressor

TEST RUNNING THE CSIR COMPRESSOR

Use the following procedure to test run your CSIR compressor. Check off the steps as you complete them. Make sure the trainer power is off while making the connections.

1. Locate the hermetic compressor on your trainer.

2. Remove the terminal cover and set it aside (it will be replaced later).

3. Locate the current relay. Find the blue wire that runs between the T_1 terminal and the only terminal on top of the current relay (see figure 8-3).

4. Cut a ⅛ inch piece of the blue wire so that current can no longer flow through it.

5. Locate the start capacitor and find the two terminals on its top.

6. Use figure 8-4 to help you complete the wiring of your compressor relay and start capacitor.

7. Connect a wire from the L_1 load side terminal in your single phase disconnect to the top terminal of the curent relays (refer to figure 8-3). Be sure to observe the proper color codes of wires.

8. Connect a wire from either terminal on the start capacitor to the top terminal on the current relay.

9. Connect a wire from the remaining start capacitor terminal to terminal 1 on the current relay.

10. Push the current relay onto the S and R terminals of the compressor.

11. Connect a neutral wire from C on the compressor (or on the overload if it has one) to the neutral terminal in the single phase disconnect. The wire from C should be connected to L_2 if the compressor is 230 or 208 volts.

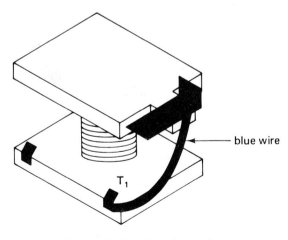

blue wire

T₁

Figure 8-3   Current relay

Test Running the CSIR Compressor

79

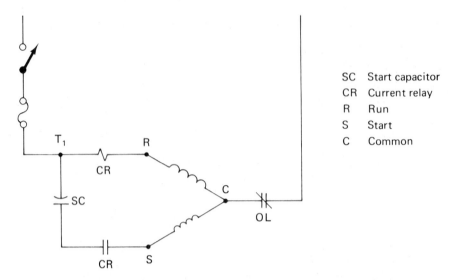

SC   Start capacitor
CR   Current relay
R   Run
S   Start
C   Common

Figure 8-4   Diagram of current relay and start capacitor

12. Place a clamp-on ammeter around the wire connecting L₁ of the disconnect to the top terminal of the current relay. Set the ammeter for 30 amp scale.

13. *Before you apply power read the complete step.* The current flow for the CSIR compressor should be very similar to the current flow in the split phase compressor, going to about 15 to 20 amps during start (LRA) and returning to about 3 to 6 amps during run (FLA). As you apply power watch the ammeter to make sure the LRA current returns to FLA in about one to two seconds. If motor does not hum or start immediately, turn off the power and call your instructor to review your wiring. With the ammeter in place, apply power to the trainer and turn the disconnect on. Be ready to turn off the disconnect if the LRA does not return to FLA.

14. Record the LRA_____

How long did it last? _____

Record FLA _____

15. Turn off the disconnect and leave all the wires in place for the next area of study.

COMPLETING A LADDER DIAGRAM FROM A WIRING DIAGRAM

As an air conditioning and refrigeration technician, you will need to be able to draw a ladder diagram from a wiring diagram to aid you in troubleshooting. Since the ladder diagram shows a sequence of operation, it is an easier diagram to use to diagnose a circuit problem. Use the following procedure to change a wiring diagram to a ladder diagram. Check off the steps as you complete them. Draw your diagram on a separate piece of paper.

1. Draw two lines three inches long each and about four inches apart (see example below).

2. At the top of each line add the labels L_1 and N if your compressor is 115 volts. Lable the lines L_1 and L_2 if your compressor is 208/230 volts.

3. Draw the symbol for a fused disconnect at the top of L_1. If your compressor is 208/230 volts, draw the disconnect switch in both lines since you used L_1 and L_2. Remember that neutral wires do not go to a switch.

4. About one inch down on L_1 draw a line across the diagram about one inch long. This line represents the wire from L_1 to the top terminal on the current relay.

5. Draw the current relay symbol so that the coil is in line with the line you have drawn in step 4.

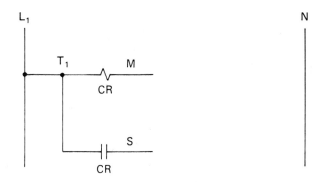

6. Draw the motor winding to the right of the current relay.

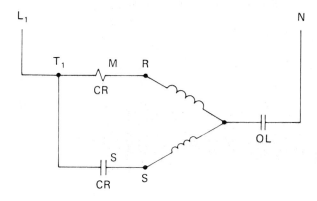

7. Draw in the wires as listed below.
   a. One wire from terminal M of the current relay to terminal R on the compressor (remember, the current relay is pushed on at this terminal so there is not really a wire, just a terminal connection).
   b. One wire from terminal S of the current relay to terminal S on the motor (again, it is really a terminal connection).
   c. One wire from the motor terminal C to N (to L_2 if it is a 208/230 volt compressor).

8. Refer to the diagram below and draw the capacitor symbol in where the blue wire was cut on your current relay.

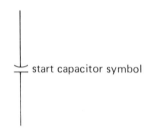

start capacitor symbol

9. Your finished diagram should look like this. Call your instructor to check your diagram.

Instructor's OK _____

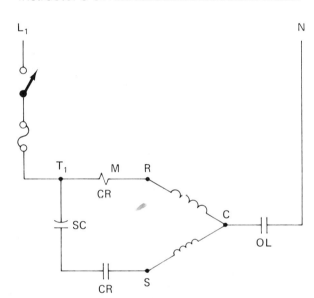

CR   Current relay
SC   Start capacitor
R   Run
S   Start
C   Common

TROUBLESHOOTING THE CSIR COMPRESSOR

Use the following procedure with the ladder diagram you have drawn to diagnose the CSIR compressor for problems. It is important that the air conditioning and refrigeration technician be able to identify problems with the CSIR compressor quickly and accurately. If the technician takes too long or makes the wrong diagnosis, the cost for getting the equipment operating will be too great, causing the customer to call another technician when the unit needs servicing again.

Ask your instructor to insert a problem into your CSIR compressor or circuit. (Possible problems include bad fuses, bad circuit wires, bad start capacitor, bad compressor, or open overload protector.)

1. Place a clamp-on ammeter around L_1 and try to make the system run. Turn the disconnect on and check the current draw as the CSIR compressor tries to start. If the current becomes excessive and the motor will not start, turn the power off immediately and go to step 2.

    a.   If the compressor runs, check the full load current to be sure it is within limits. If the LRA is within limits you have completed this test. Ask your instructor to put a new problem into your system.

2.   If the compressor tries to run (makes a noise and draws high current) but does not start, skip to step 4.

3.   If the compressor does not try to run (does not make a noise and current draw is zero), skip to step 5.

4.   Since your compressor is trying to start it means that the circuit wires are probably functional. The problem is therefore the current relay, start capacitor, or compressor motor windings. Make the following tests in order:

    a.   Place the ammeter around the wire that connects the start capacitor to the current relay. This time apply power to the system for two seconds and shut the power off. Watch for LRA current draw in the start winding. If the current is zero skip to step 4b. If the current is more than 1 amp, skip to step 4c.

    b.   Since the current in the start winding in step 4a was zero, the problem is in the current relay, start capacitor, or start winding. Look at figure 8-5 and your ladder diagram, and notice that these parts make a series circuit. There must be an open in your CSIR circuit since there is no current in the L_1 line (test step 2). There is complete current at least from L_1 to the run winding and back through the motor terminal C to N (see figure 8-5).

        Set your voltmeter to the 250 volt scale. Place one probe on the start terminal that has a wire connected to terminal T, and one probe on the N terminal of the disconnect. This time turn the power on for two seconds and immediately shut it off. Watch the voltmeter to see if you have power to this point.

    Record the amount of voltage _____

    c.   If the voltage at this point is zero, check your start capacitor and the two wires connected to your capacitor. If the capacitor is bad, ask

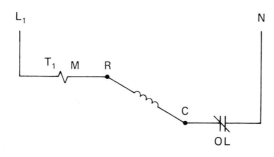

Figure 8–5

Trouble-shooting the CSIR Compressor

your instructor for a replacement capacitor and return to step 1. If the wire is bad, repair it and return to step 1.

d. If the voltage at this point is full applied voltage, the trouble is in the contacts of the current relay. Test your current relay with the procedure in Unit 7. Ask your instructor for a replacement part if needed. If the current relay is good, check the amount of resistance from terminal S to C in your compressor; refer to Unit 6 if you need help. Replace parts as needed and return to step 1.

e. Since the current in step 4a is more than 1 amp in the start winding circuit, you can assume it is functioning correctly. The problem is in the run winding. Check the run winding according to the procedure in Unit 6. Replace parts as required and repeat this test starting with step 1.

5. Since the compressor did not try to start and current flow through L$_1$ was zero, the problem is a loss of voltage either at the disconnect or somewhere in the circuit wiring or components (see figure 8-6).

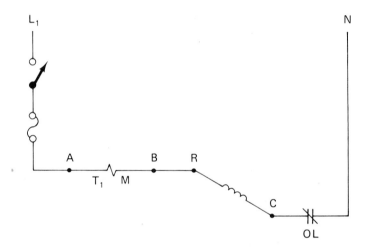

Figure 8–6

Use a voltmeter to find where the voltage stopped. Check the next steps in order.

a. Leave the ammeter in place on L$_1$ and apply power again to the compressor circuit. Be sure there is no current flow at all. Since there is no current flow, you can leave the power on as long as it takes to find the problem. Check the single phase disconnect for proper voltage on the load side terminals and record volts _____. If the voltage is zero, check the line-side terminals.

b. If the voltage is still zero, you have found the problem. Ask the instructor to return power to your disconnect.

c. If the voltage is full applied voltage at the line side terminals, check the fuses. Replace parts as required and repeat step 1.

d. If the voltage at the load side terminal is full applied voltage, go to step e.

e. Since you have voltage at the disconnect but L_1 shows no current, the problem has to be an open wire or bad component. Use the same technique as described in Unit 7 to find the loss of voltage. Leave one probe at the neutral terminal in the disconnect and use the other probe to test the following points on figure 8-6:

Record the voltage at point A _____

point B _____

f. If the voltage at point A is zero, the wire between point A and the disconnects is bad. Replace it and return to step 1.

g. If the voltage at point A is full applied voltage, continue to step h.

h. Check voltage at point B. If voltage at point B is zero, the problem is the current relay. Check the current relay according to the procedure in Unit 7. Replace parts as needed and go to step 1. If voltage is still full applied, go to step c.

i. Reverse the probes: put one probe on terminal L_1 in the disconnect, and move the other probe to the terminal on the compressor overload. If the voltage is zero, the wire from N to the compressor overload is bad. Replace the wire and return to step 1. If voltage is still full applied, go to step 5.

j. Leave one probe on L_1 in the disconnect and move the other probe to terminal C on the compressor. If the voltage is zero the compressor overload is open. Check its temperature. If it is cold but will not pass current, it is defective. Replace the overload and return to step 1. If the compressor overload is warm to the touch, allow it to cool and return to step 5j. If voltage is still zero, replace the overload and return to step 1. If the voltage is still full applied voltage, continue to step k.

k. The voltage across the compressor terminal is full applied in step 5j. You must assume the compressor is bad. Check the compressor, following the procedures in Unit 6. Replace parts as needed and return to step 1.

You have now completed the troubleshooting sequence for CSIR compressors. Turn the power to your trainer off and remove all wires. When you are ready for the quiz on Unit 8, call your instructor.

9 The Permanent Split Capacitor Compressor

Tools Required

Voltmeter (0-250 volt scale)
Screwdriver (flat blade ¼ in.)
Clamp-on (30 amp capacity)

Objectives for Unit 9

At the conclusion of this lesson the student should be able:

To distinguish a run capacitor a from start capacitor.

To draw accurately the electrical diagram for a run capacitor.

To explain orally or in writing to the instructor's satisfaction the reason a run capacitor has one terminal marked with a red dot.

To explain orally or in writing to the instructor's satisfaction the theory of operation of a permanent split capacitor compressor.

Given a wiring diagram of a permanent split capacitor compressor, to wire accurately and test run the compressor.

Given a wiring diagram of a permanent split capacitor compressor controlled by a low voltage relay and thermostat, to wire accurately and run the system.

To change the wiring diagram in objective 5 into a ladder diagram.

THE PERMANENT SPLIT CAPACITOR COMPRESSOR

The current relay we discussed in Unit 7 can only be used on small compressors, usually those of less than one horsepower, because a larger current is required to run the bigger compressors. This current must travel through the coil of the current relay (review Unit 7). The current relay coil is not able to handle the load on larger compressors and burns open more frequently due to this large current flow.

For this reason, manufacturers have found a more reliable method of powering the start winding (see figure 9-1) so the compressor can still get the effects of the phase shift caused by the capacitor. For this compressor the capacitor is placed in a steel case instead of plastic, and is permanently left in series with the start winding. This is why this wiring configuration is called a *permanent split capacitor* compressor.

The permanent split capacitor compressor, also called PSC, has become the most widely used compressor in air conditioners and refrigerators. Since there is no relay to disconnect the start winding after starting, there is one less component to fail. In the PSC compressor, the capacitor is the only component used to start the compressor.

TESTING THE RUN CAPACITOR

Since the capacitor is left in the start winding circuit the entire time the compressor is running, it is called a *run capacitor* (see figure 9-2).

Because the run capacitor is always in a steel case and the start capacitor is in a plastic case, they are easy to tell apart. The steel case allows the heat built up due to the long run cycle to be dissipated to the air.

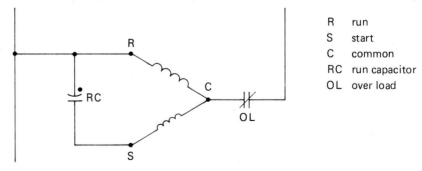

R   run
S   start
C   common
RC   run capacitor
OL   over load

Figure 9-1   Diagram of permanent split capacitor compressor

In figure 9-3 you will notice that the electrical symbol for the run capacitor is similar to that for the start capacitor. The difference is that the run capacitor symbol is identified by the letters RC and the start capacitor uses the letters SC.

You should also notice that one terminal of the run capacitor is marked with a dot. This dot will be either red paint or the manufacturer's symbol. The dot represents the terminal that is connected to the capacitor plate which is closest to the capacitor's steel case. Remember from Unit 8 that a capacitor is composed of two plates (sheets of foil) that are separated by an insulator (sheet of dielectric impregnated paper). These layers are rolled up like a rug (see figure 9-4) and placed into the steel can. This means that one of the foil plates will be very close to the steel can, separated only by the insulating paper. If the insulation shorts through the capacitor, it will draw excessive current.

Figure 9-2   Run capacitor (courtesy of *General Electric*)

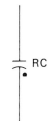 RC

Figure 9-3   Electrical symbol for run capacitor

Figure 9-4   Rolled capacitor

If the red dot terminal is connected to the L₁ of the disconnect, this short will only cause a fuse to blow. If the red dot terminal is connected to the start winding of the compressor, the short will burn out the compressor start winding. For this reason, *always be sure the "red dot" terminal is wired to the disconnect and not the compressor start winding.*

Some run capacitors have three terminals, marked H,C, and F. This type is really two capacitors in one container. One capacitor is found between terminals C and H (C for common, H for hermetic) and is the capacitor for the compressor. The other capacitor also uses the common terminal and terminal F (fan), and is used for a fan motor.

The run capacitor is tested for opens and shorts the same way as the start capacitor. Locate the run capacitor on your trainer. Use the procedure in Unit 8 to test your run capacitor.

PSC MOTOR THEORY

Since the capacitor is in series with the start winding, the capacitance phase shift during start will be similar to the capacitor start induction run compressor. The run capacitor is usually rated between 5 uf and 50 uf. Since this rating is smaller than that of the start capacitor, the PSC motor will have a little less starting torque than the CSIR compressor, but more than the split phase compressor studied in Unit 7.

Once the PSC compressor starts and begins to run, the motor will again produce counter-electromotive force. This CEMF will build up to within a few volts of the applied voltage when the compressor is at full speed. As long as the difference between the applied voltage and the CEMF is small, very little current will flow in the start winding. This is because the capacitor will allow more current to pass as the difference in applied voltage and CEMF gets larger, and less current to flow when the voltage difference is small. Since the voltage difference at full RPM is small, the current in the start winding will be small, approximately 2 to 4 amps. This small current will not be enough to damage the compressor start winding.

Another effect of leaving the run capacitor in the start winding during the run cycle is that it helps to regulate the compressor speed. If the compressor starts to load up (more refrigerant needs to be pumped), the compressor speed will tend to slow down slightly. When the RPM slows down the CEMF becomes less, making a bigger voltage difference. This causes the capacitor to allow more current to flow in the start winding. This current gives the motor a little more torque, allowing the compressor to move the heavier load. As the load passes, the compressor again runs at full RPM and the current in the start winding returns to the small amount.

TEST RUNNING THE PSC COMPRESSOR

Use the following procedure to test run the PSC compressor. Check off the steps as you complete them.

1. Turn the power to your trainer off.

2. Locate the hermetic compressor on your trainer and record the following data from the name plate.

Name of manufacturer _____

Model #   _____

Serial #   _____

Volts   _____

LRA   _____

Hertz   _____

3. Remove the terminal cover and set it aside. It will be replaced later.

4. Draw a sketch of the three compressor terminals showing the accurate location of each terminal.

5. Use the procedure outlined in Unit 6 to identify the three terminals as *start*, *run*, and *common*. Use an R, S, or C to properly identify the terminal in your sketch.

6. Use this diagram (figure 9-5) to wire your compressor as a PSC compressor.

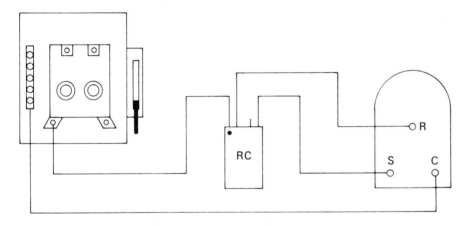

Figure 9–5

a. Connect a wire from the L_1 load-side terminal in the single phase disconnect to the terminal of the run capacitor identified by a red dot. The red dot identifies the plate of the capacitor that is closest to the steel container and must be wired to terminal L_1.

b. Connect a wire from the same "red dot terminal" to terminal R on the compressor. Notice that each terminal of the run capacitor has room for two wires.

c. Connect a wire from the second capacitor terminal to the start winding of the compressor.

d. Connect a wire from the common terminal of the compressor to the neutral bar in the single phase disconnect. Use L_2 in the single

Test Running the PSC Compressor

91

phase disconnect instead of neutral if the compressor is 230/208 volts.

7. Place the clamp-on ammeter on the wire that connects terminal S on the compressor to the capacitor. Set the ammeter range to 30 amps.

8. *Read the following statement completely before you turn the power on:* When the PSC compressor starts, the current in the start winding will increase to about 15 to 20 amps. Once the compressor begins to run (usually two to three seconds), the current in the start winding will return to approximately less than 5 amps. The current in the run winding (FLA) may be as high as 12 to 20 amps depending on the horsepower rating of the compressor. To be sure the PSC compressor is operating correctly, first check the start current. Make sure it returns to below 5 amps once the compressor starts. If the current in the start winding stays high, *immediately* turn the power off and call your instructor. If the current of the start winding returns to normal, use your ammeter to measure the FLA at the wire connected to the run winding. If this current is more than 20 amps, *immediately* turn the power off and call your instructor. With the ammeter in place on the start wire, observe the precautions outlined above while turning on the power.

9. Record current in the start winding when the compressor is running: _____ amps.

10. Record the current in the run winding while the compressor is running: _____ amps.

11. After you have made the measurements in steps 9 and 10, turn the disconnect off and repeat step 8 once more to familiarize yourself with the current draw of the PSC compressor during start and run.

12. Turn the power to your trainer off. Leave the wires in place and continue to the next section.

CONTROLLING THE PSC COMPRESSOR

Since the permanent split capacitor compressor is usually larger than the split phase and CSIR compressors, a relay is usually used to turn it off and on. You may need to review Unit 4 and Unit 5 to remember how to wire a 24 volt transformer, a 24 volt relay, and a low voltage cooling thermostat.

In this exercise the PSC compressor will be controlled by the relay contacts. Use the following steps to control the PSC compressor with a 24 volt relay. Follow the diagram in figure 9-6 and check off the steps as you complete them:

1. Be sure the power to your trainer is turned off.

2. Locate the 24 volt relay that has two sets of normally open (N.O.) contact, the control transformer, and low voltage thermostat.

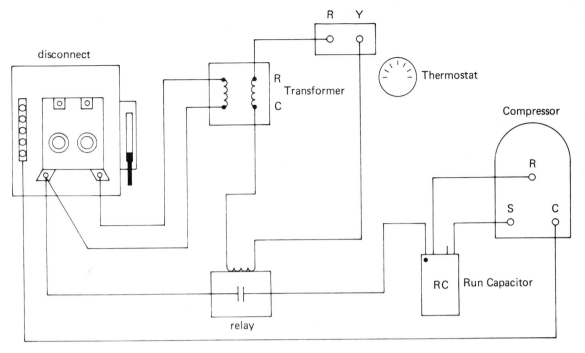

Figure 9-6

3. Connect this part of the circuit as you did in Unit 5. Be sure to use the proper primary voltage to your control transformer. Refer to figure 9-6 as you complete the low voltage wiring.

4. Remove the wire from terminal L_1 in the disconnect and connect it to the load side of the left set of relay contacts.

5. Connect a wire from the line side of the same contact to the load side of T_1 in the disconnect. If your compressor is 115 volts, skip steps 6 and 7. If the compressor is 208/230 volts, complete steps 6 and 7.

6. Remove the wire from terminal L_2 of the disconnect and reconnect it to the load-side terminal on the right side of the relay contacts.

7. Connect a wire from the load-side terminal L_2 of the disconnect to the line-side terminal of the right-hand set of contacts.

8. Place the ammeter around the wire from the capacitor to the start winding and set the range to 30 amps.

9. Set the thermostat switch to cooling and the lowest temperature.

10. Observe the same precautions as you did in step 8 in the run procedure during the starting of the PSC compressor. If the current goes too high or the compressor fails to start, turn the power off at the disconnect *immediately* and call your instructor. Turn the disconnect on at this time.

11. After the compressor has run for two or three minutes, turn the thermostat to the highest number. This should open the relay contacts

Controlling the PSC Compressor

93

and turn the compressor off. If the compressor will not shut off, call your instructor to inspect your circuit.

12. Wait two minutes and turn the compressor to the lowest number again. Be sure to observe all precautions outlined above during the starting process. The PSC compressor should start again after the contacts have closed.

13. This completes the control test. Turn the power off at your trainer and remove all wires. Place all components in their proper locations.

CONVERTING THE WIRING DIAGRAM TO A LADDER DIAGRAM

At times it will be important for the air conditioning technician to change a wiring diagram to a ladder diagram to diagnose problems in the electrical system. The easiest way to draw the ladder diagram is to start with the two power supplies. The first will be the applied line voltage. The second will be the 24 volt control circuit. These are the supply voltages that will be provided for your ladder diagram in figure 9-7. Use this space to convert the wiring diagram in figure 9-6 to a ladder diagram. Follow the instructions listed below.

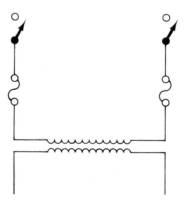

Figure 9–7

Complete the high voltage portion of the circuit first. This will include the PSC compressor, run capacitor, and relay contacts. Draw these in between the two lines representing the applied voltage. Refer to the ladder diagrams for the CSIR and split phase compressor if you need help.

The control voltage portion of this diagram will include the cooling thermostat (terminals R and Y), the relay coil, and terminals R and C of the control transformer. Review Unit 5 if you need help. Call your instructor for approval when you are finished. Instructor's approval _____

This completes Unit 9. Review the objectives for Unit 9 and be sure you can complete them. Call your instructor when you are ready to take the quiz on Unit 9.

10 The Capacitor-Start, Capacitor-Run Compressor

Safety for Unit 10

At times the circuits and disconnects you will be working with will be powered with up to 230 volts. You must be aware of and take appropriate safety precautions relative to electrical shock hazards and wear safety glasses while working. Start and run capacitors will be used during unit. You should follow the instructions provided in Unit 9 to handle and discharge capacitors.

Tools Required

Voltmeter (0-250 volt scale)

Screwdriver (flat blade ¼ in.)

Clamp-on ammeter (30 amp capacity)

Objectives for Unit 10

At the conclusion of this lesson the student should be able:

To draw accurately a diagram of a potential relay and identify the terminals.

To explain either orally or in writing the theory of operation of a capacitor-start, capacitor-run compressor, including the start capacitor, run capacitor, and potential relay.

To draw accurately the ladder diagram and wiring diagram of a capacitor-start, capacitor-run compressor.

Given the diagram in objective 3, to wire correctly the start capacitor, run capacitor, and potential relay to the compressor, and test run the compressor.

To add a control circuit to a capacitor-start, capacitor-run compressor and test run the compressor.

To find a given electrical problem in the electrical circuit of objective 5 to the instructor's satisfaction.

CAPACITOR-START, CAPACITOR-RUN COMPRESSOR

Some air conditioning and refrigeration compressors require very large starting torque. To give the single phase compressor the most starting torque, both a start capacitor and a run capacitor are used. Since the compressor uses both capacitors it is called a capacitor-start, capacitor-run, or CSCR, compressor.

Equipment manufacturers will generally use the CSCR compressor on refrigeration equipment that uses an expansion valve and on air conditioning systems when the permanent split capacitor (PSC) compressor has trouble starting. In some cases the start capacitor is merely added in the field to the PSC compressor after it has demonstrated problems with starting. When the start capacitor is added in the field it is called a *hard start kit*.

TESTING THE POTENTIAL RELAY

Since the CSCR compressor adds a start capacitor to the run capacitor, a method to remove the start capacitor from the start winding must be used. The current relay will not work because of the large currents the CSCR compressor uses. Another relay, called the *potential relay*, will be used (see figure 10-1).

The potential relay has a high resistance coil (approximately 10K-20K ohms) that is energized by the counter-electromotive force (CEMF) produced by the compressor. The symbol (see in figure 10-2) for the potential relay shows the coil between terminals 2 and 5 of the relay and the normally closed contacts between terminals 2 and 1. Use the following procedure to test the potential relay with an ohmmeter. Check the steps off as you complete them.

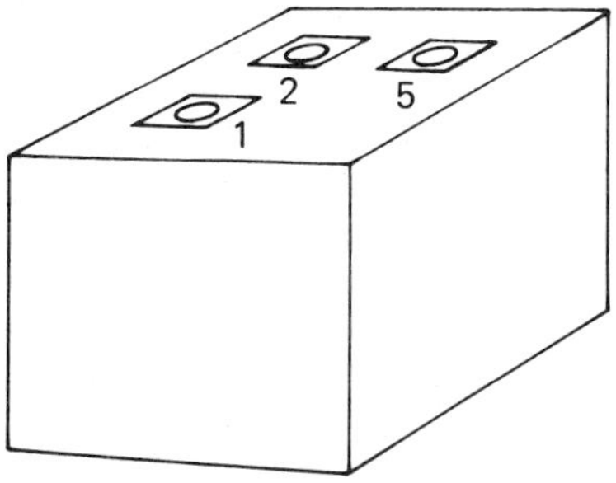

Figure 10-1   Potential relay

Figure 10–2

1. Locate the potential relay on your trainer. Remove the mounting screws and take the potential relay from the trainer. Replace the mounting screws on the trainer so they won't be lost.

2. Draw in the terminals and their numbers in the outline of the potential relay below.

3. Set the ohmmeter to 100K scale and zero.

4. Place the meter probes on terminals 2 and 5. This is the coil, and its resistance should be approximately 10K-20K ohms.

    Record the resistance of your coil _____

    a. If your reading is infinity (∞) on the R $\times$ 100K scale, the coil has developed an "open" and the potential relay must be replaced.

    b. The resistance reading may be a little lower than 10K. If it is much below 10K you do not have a potential relay. It is a control relay and will be used in later control circuits. It must be pointed out at this time that relay manufacturers use the same plastic case for potential relays and regular control relays. Therefore, it is essential that you check the relay coil. If the coil resistance is betwen 10K and 20K ohms, it is a potential relay. If it has less resistance, it is most likely a control relay.

5. Reset the meter to R $\times$ 1 and zero.

6. Place the meter probes on terminals 1 and 2 of the relay. These terminals are connected to the normally closed contacts of the relay and the resistance reading should be less than 1 ohm.

    a. If the measurement is less than 1 ohm the relay contacts are good. Remount the relay on your trainer and continue to the next section.

    b. If the resistance measurement is more than 1 ohm, the contacts of your relay are dirty or pitted and the relay should be replaced. Ask your instructor for a new potential relay and repeat this test starting at step 1.

This completes the test of your potential relay with an ohmmeter. This type of test is called a static test since the relay did not really operate. It is used because it is quick and reliable. A dynamic test (in operation) can be made when the potential relay is used on the CSCR compressor. A dynamic test will be made later in this unit.

THEORY OF OPERATION FOR THE CSCR COMPRESSOR

The CSCR compressor has the start capacitor in parallel with the run capacitor. Both capacitors are in series with the start winding (see figure 10-3). Connecting capacitors in parallel adds their capacitive values. The formula $C_T = C_1 + C_2$ is used for adding parallel capacitors—similar to resistors in series.

The effect of having a start capacitor with a large capacitance value (up to 600 uf) added to the value of the run capacitor (up to 75 uf) will cause a larger phase shift between the start and run winding voltages. Referring to figure 10-3, you will notice that during start the potential relay contacts are normally closed, completing the circuit between the start capacitor and the start winding. As the compressor motor starts to turn, the CEMF begins to build. The CEMF will be present at terminals S and C. The potential relay coil (terminals 2 and 5) is connected to terminals S and C on the compressor. As the compressor reaches approximately 75% of full RPM, the CEMF will be strong enough to energize the potential relay coil and pull its contacts open. When the potential relay contacts open, the start capacitor is removed from the start winding.

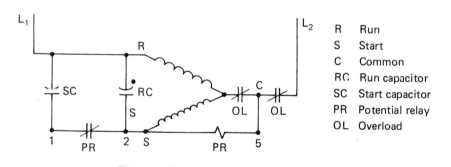

Figure 10-3   CSCR compressor

The run capacitor is left in the start winding circuit, as in the PSC compressor. This gives the CSCR compressor excellent starting torque and good running efficiency.

TEST RUNNING THE CSCR COMPRESSOR

Use the following procedure and figure 10-4 to connect the potential relay, start capacitor, and run capacitor to your compressor, and test run the compressor. Observe the safety precautions and check off the steps as you complete them.

1. Turn off all power to your trainer.

2. Locate the start capacitor, run capacitor, and potential relay on your trainer.

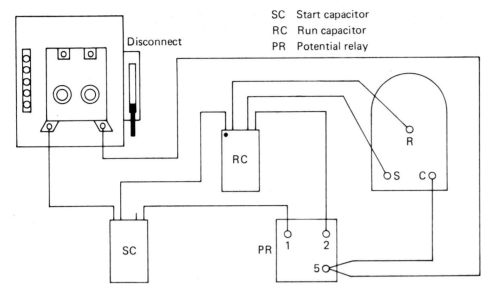

SC   Start capacitor
RC   Run capacitor
PR   Potential relay

Figure 10-4   CSCR compressor

3. Locate the compressor on your trainer and record the following data. You will use this data to identify the compressor.

    Volts _____   Model #_____

    Hertz _____   Serial # _____

    LRA _____

4. Connect a wire from load-side terminal L_1 on your single phase disconnect to one side of the start capacitor (black plastic case).

5. Connect a wire from the same terminal of the start capacitor (it has room for two wires) to the "red dot" terminal of the run capacitor (steel case).

6. Connect a wire from the same "red dot" terminal on the run capacitor (it also has room for two wires) to the R terminal of the compressor.

7. Connect a wire from the other terminal on the start capacitor to terminal 1 on the potential relay.

8. Connect a wire from the second terminal on the run capacitor to terminal 2 on the potential relay.

9. Connect a wire from the same terminal of the run capacitor you used in step 7 to terminal S on the compressor.

10. Connect a wire from terminal 5 on the potential relay to terminal C on the compressor.

11. Connect a wire from terminal 5 on the potential relay to the neutral terminal in the single phase disconnect (L_2 if the compressor is 208/230 volts).

12. Call your instructor to inspect your wiring at this time.

    Instructor's approval _____

13. Place your clamp-on ammeter around the wire that goes from the run capacitor to terminal S on the compressor and set the scale to 30 amps.

14. *Read this step completely before you turn on the power*. The potential relay should allow the start capacitor to provide a phase shift to help start the compressor, and then completely remove the start capacitor from the circuit. The run capacitor will stay in the circuit for the entire time the compressor is running. You will observe the ammeter reading to tell if the start capacitor is operating correctly. The ammeter should read approximately 20 amps for about 1 to 2 seconds during start, then the current should drop to less than 5 amps showing the start capacitor is out of the circuit and the run capacitor is operating correctly. If the current stays at zero or does not return to less than 5 amps turn the power off immediately and call your instructor to check your system.

15. Turn the power on at the disconnect at this time. Observe the current with the ammeter and record the current at S while the compressor is running _____.

16. While the compressor is still running move the ammeter to the wire that connects the L_1 terminal of the disconnect to the start capacitor. The current will be the full load amperage for your compressor.

    Record the FLA _____.

17. Turn the disconnect off. You will not need to wait to restart the CSCR compressor as you did for the PSC compressor. The high starting torque of the CSCR compressor will allow it to restart almost immediately after it has been turned off. Restart your compressor by repeating step 15. If the compressor will not restart, shut the power off immediately and call your instructor to check your system.

18. Turn the disconnect off at this time. Leave all wires and components connected for the troubleshooting and control exercise.

CONTROLLING THE CSCR COMPRESSOR

Since it uses a large current to start and run, the CSCR compressor will normally be controlled by a relay. In air conditioning and refrigeration systems a low voltage thermostat is used to turn the relay coil off and on. Figure 10-5 shows the CSCR compressor controlled by the low voltage thermostat and relay.

Use figure 10-5 and refer to Unit 9 to connect the relay and thermostat to control the CSCR compressor. The following procedure will aid you. Check off the steps as you complete them.

1. Turn off the power to the trainer.

2. Connect the low voltage portion of the circuit as shown in figure 10-5. Refer to Unit 9, steps 1 to 3, on controlling the PSC compressor if you need help.

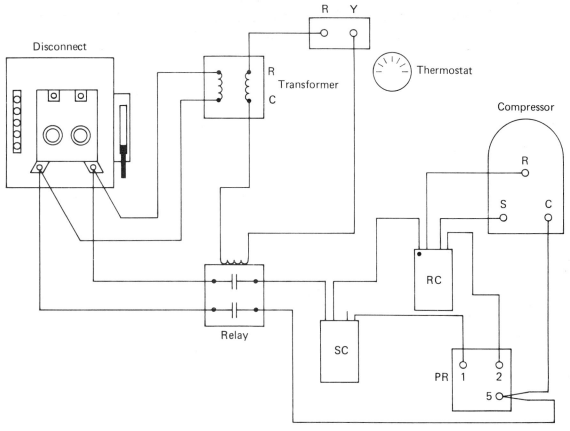

Figure 10-5 Diagram of low voltage thermostat controlling CSCR compressor

3. Disconnect the wires from terminal L_1 and L_2 at the load side of your disconnect. Reconnect these wires to the load-side terminals of your relay. Be sure you are using the relay that has a 24 volt coil.

4. Add wires from the load side terminals of L_1 and L_2 of your disconnect to the line-side terminals of your relay.

5. Call your instructor to inspect your wiring.

Instructor's approval _____

6. Place your clamp-on ammeter around the wire going to the S terminal on the compressor and follow the starting instructions provided in step 15 of the procedure to test run the CSCR compressor. Turn the thermostat to "cooling" and set the dial to the lowest temperature. Turn on the disconnect to power your circuit and observe the current reading as instructed in step 15 of the test run procedure.

Record the full load amperage (FLA) _____

7. Cycle the compressor off by setting the thermostat to the highest temperature. If the system will not turn off, call your instructor.

8. Cycle the compressor back on after 15 to 20 seconds. Again notice that the CSCR compressor has good starting torque and can easily restart.

9. Turn the power off at the disconnect and leave all wires in place for the troubleshooting procedure.

DRAWING A LADDER DIAGRAM OF THE CSCR

Before you can troubleshoot the CSCR compressor, you must have a good ladder diagram of the system. Often there will be no ladder diagram available, so that you must develop one from the wiring diagram or by following the wires on the sysem. Use the following procedure to draw the ladder diagram for the CSCR compressor. If you need help, refer to any of the previous units where ladder diagrams have been discussed.

1. The line voltage and low voltage have been drawn in for you. Be sure to label the amount of your line and low voltage on the diagram.

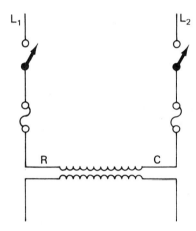

2. Draw a diagram of the CSCR compressor (figure 10-3) between L_1 and L_2.

3. Draw the two sets of normally open relay contacts, one between L_1 and the line showing the wire going to the run capacitor and start capacitor, the other between L_2 and the compressor overload.

4. Draw the low voltage circuit identical to the ladder diagram you made in Unit 9.

5. Call your instructor to approve your ladder diagram.

Instructor's approval _____

TROUBLESHOOTING THE CSCR COMPRESSOR AND CONTROL CIRCUIT

In order to find electrical problems in the CSCR circuit quickly and accurately, a combination of observations and electrical measurements will be used.

Use the ladder diagram you have just developed to identify the points to make your measurements. Follow this procedure and check off the steps as you complete them. Ask your instructor to put a problem in your circuit. The problems that can be in the following components or circuits, (just as you would find them in the field) will be: fuses, transformers, thermostat, relay, start capacitor, run capacitor, potential relay, compressor interconnecting wires, or loss of voltage.

1. Set the thermostat on cooling and to the lowest temperature. Try to make the system run. Put your ammeter in place and observe the procedure outlined in making the CSCR compressor run.

   a. If the compressor makes a noise but does not start, immediately turn the disconnect off and skip to *step 2*.

   b. If the compressor will not make a noise and does not try to start, turn the power off at the disconnect and skip to step 5.

   c. If the compressor runs, check the FLA. If the LRA returns to the full load (FLA) level after starting, the compressor is considered to be operating correctly. Ask your instructor for another problem if you need more practice.

2. Since the compressor tried to run, you must assume the control circuit has made the relay contacts closed. This time, closely observe the ammeter to see if any current is flowing through the start winding. Remember to turn the disconnect on for only one to two seconds so as not to overheat the overload or motor windings.

   a. If LRA current is present, skip to step 3.

   b. If no current is present, skip to step 4.

3. Since current is flowing in the start winding, the problem is in the run winding. Put one probe of your voltmeter on L_2 in the disconnect and the other on R at the compressor. Turn the compressor on for one to two seconds and check for voltage.

   a. *Voltage is present*. Test the compressor run winding for continuity. Replace parts as needed and return to step 1.

   b. *No voltage is present*. Repeat the voltage test, following the wiring back to terminal L_1. Since the start winding shows current draw, the problem is a break in the circuit (or component) between the point where the start capacitor gets power and the R terminal of the compressor. Make repairs and return to step 1.

4. Since there is no current through the start winding, you must assume there is an open somewhere between terminal L_1 in the disconnect through the start capacitor, potential relay, and start winding. Leave one voltmeter probe on L_2 and move the other from L_1 to the point where voltage ends. This point indicates you have just passed the open. Back up one test point and make repairs in the wiring, replacing components as needed. Remember to allow the motor to try to run for only one to two seconds during each test. Do not overheat the motor or allow an overload to develop. Return to step 1 when repairs have been made.

5. Since the compressor did not make a noise or try to run, you must assume that either the control circuit or the line voltage circuit has a loss of power at the source, in the wiring, or in a component. To find the part of the system where the voltage has stopped, make the following tests that will tell where the open circuit is. Try to start the compressor again and watch to see if the relay pulls its contacts closed.

   a. Relay contacts "click" closed. Skip to step 6.

   b. Relay contacts do not move. Skip to step 7.

6. Since the contacts closed, you have proved the control circuit is operating correctly. This means the transformer, cooling thermostat, and relay are all good and are working correctly. Repeat step 4 in this procedure. You are looking for a break in the circuit caused by a broken wire or bad component in the line voltage circuit. Use the voltage loss method of testing. Leave one probe on L_2 of the disconnect and move the other probe from point to point until you find where voltage stops. You can leave the power on all the time you are testing since no current is flowing. Make repairs or replacements as necessary and return to step 1.

7. Since the relay contacts have not pulled closed, you must assume the problem is either a loss of voltage (blown fuse) or in the low voltage circuit.

   a. Test for voltage at L_1 and L_2 in the disconnect. Replace the fuse if needed and return to step 1. If voltage is present go to step 7b.

   b. Since voltage is present at L_1 and L_2, test for low voltage at R and C. If 24 volts is present skip to step 7c. If no voltage is present, test the transformer as instructed in Unit 4. Replace parts as needed and return to step 1.

   c. Since 24 volts is present at R and C on the transformer, leave one voltmeter probe on terminal C of the transformer and move the other probe to the following points: R of thermostat, Y of thermostat, and then to the relay coil. You are looking for a loss of voltage just as you were in the line voltage circuit. At the test point where voltage is zero, back up one test point; the problem is between these two points. Make appropriate repairs and return to step 1 of this procedure.

After finding the electrical problem, you have completed the trouble-shooting procedure. Remember that this procedure should be used any time in the lab or on the job when you are trying to find an electrical problem.

This completes Unit 10. Call your instructor when you are ready to take the quiz on Unit 10.

11 Split-Phase Open-Type Motor

Safety for Unit 11

At times the circuits and disconnects you will be working with will be powered with up to 230 volts. You must be aware of and take appropriate safety precautions relative to electrical shock hazards and wear safety glasses while working. You will also be operating a motor with an open rotating shaft. You must not wear loose clothing and must keep hands clear of this shaft when it is turning.

Tools Required

Voltmeter (0-250 volt scale)

Screwdriver (flat blade ¼ in.)

Clamp-on ammeter (30 amp capacity)

Needle nose pliers (3 in. with insulated handles)

Objectives for Unit 11

At the conclusion of this lesson the student should be able:

To explain to the instructor's satisfaction the theory of operation of the split-phase open-type motor.

To identify correctly either orally or in writing three uses for open-type motors in air conditioning or refrigeration systems.

To wire correctly the split-phase open-type motor for a change of voltage, and test run on 115 volts and 230 volts.

To wire correctly the split-phase open-type motor for a change of rotation, and test run the unit.

To diagnose accurately electrical and mechanical malfunctions in the split-base open-type motor and in its electrical system.

SPLIT-PHASE OPEN-TYPE MOTORS

Most air conditioning and refrigeration systems need fans to move air across their evaporators and condensers. Some systems need a pump to circulate water. All of these applications require a motor that has the shaft directly accessible to attach or mount pulleys, gears, blades, or pumps. This type of motor is called an *open type* motor.

This open-type motor (see figure 11-1) has the shaft protruding through one end plate. Unlike the hermetic motor, the open motor can have some minor changes made in the field while it is on the equipment. These changes may be necessary to change the directions of the shaft's rotation, the speed of the shaft (RPM), or the voltage requirement (for example, from 230 to 115 volts).

Not all open motors can have their rotation, voltages, and speeds changed. This capability must be built into the motor when the motor is manufactured. The reason air conditioning technicians need to know how to change rotation, speed and voltage is to allow them to carry fewer motors in their truck or shop as replacement parts. If you have one motor that can be wired for 230 or 110 volts, clockwise or counterclockwise rotation, and 1150 or 1725 RPM, you will not have to stock eight different motors to fit all these combinations. Remember, you cannot predict what motors you will need when you get a service call. You must be prepared to go to the site, find the problems, and install new parts in one trip.

Open motors cannot have their starting torques changed like hermetic compressors can. Instead, you must stock three separate motors for

Figure 11-1 Open-type motor
(courtesy of *General Electric*)

different starting torque requirements: the permanent split capacitor (PSC), the split phase, and the capacitor start induction run (CSIR). (The capacitor start capacitor run is not generally used as an open type motor.) This unit will deal only with the split-phase open motor. The next units will explain typical uses for these motors, along with installation and troubleshooting procedures.

THEORY OF OPERATION OF THE SPLIT-PHASE OPEN MOTOR

The split-phase open motor operates in a manner very similar to the split phase compressor. It still has a run winding and a start winding. The theory of operation is essentially the same as for hermetic motors. The start winding must still be removed from the circuit as soon as the motor is started. In open type motors, a device called an *end switch* (see figure 11-2) disconnects the start winding from the circuit at approximately 75% full RPM.

Notice in figure 11-3 that the end switch has two movable weights mounted on the shaft of the rotor. As the shaft spins, the weights are pulled outward by centrifugal force. At 75% full RPM their movement is enough to move the spool to the right and allow the end switch contacts to open.

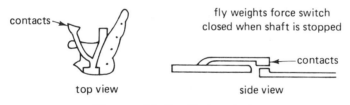

Figure 11-2   End switch

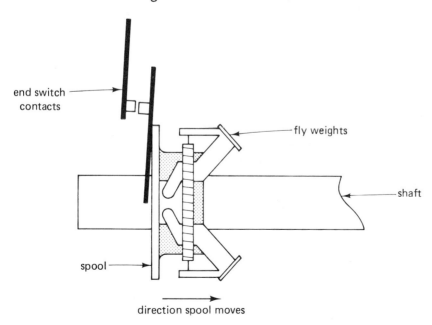

Figure 11-3   End switch and fly weights

The movement of the end switch and weights will cause a distinct click, once as the motor starts and again when the motor is slowing to a stop. You will be asked to identify this click later.

TEST RUNNING THE SPLIT PHASE OPEN MOTOR

Use the following procedure to connect the wires to the split phase motor and test run it. Check off the steps as you complete them.

1. Turn off all power to your trainer.

2. Locate the split-phase open motor and record the following data from the data plate.

Model # _____   S.F. _____

Serial # _____   Rotation _____

Voltage _____   RPM _____

Amps _____   Frequency _____

H.P. _____   Phase _____

H.P. Horsepower rating. This tells how big a load the motor can turn.
S.F. Service factor rating. This tells by what percentage the motor horsepower rating can safely be exceeded or overloaded.
Rotation. This will be listed as CW (clockwise) or CCW (counter-clockwise).
RPM. The speed of the shaft.

3. Remove the service and inspection plate on the end of the motor (see figure 11-4).

     Notice!!   Lay the inspection plate cover where it will stay clean. It may have some necessary diagrams in it, so be careful not to lose it. It will be replaced when you have completed wiring the motor.

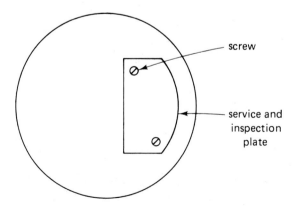

Figure 11-4   Motor inspection plate

4. Locate the terminals marked L₁ and L₂. On some motors these terminals may be marked with just an L. In most cases these terminals will be the only two that are the nut-and-bolt type.

5. Connect a wire from loadside terminal L₁ of your single phase disconnect to terminal L₁ on the motor terminal. Be sure to bring the end of the motor terminal wire through the access hole in the end plate, and not through the inspection cover.

6. If the motor is 230 volts, connect the wire from terminal L₂ in the disconnect to terminal L₂ on the motor terminal board. If the motor is 115 volts, use the neutral terminal in the disconnect instead of L₂. *NOTE:* If the motor data plate identifies the voltage as 230/115, call your instructor to determine what voltage the motor is wired for at this time. This procedure will be explained to you later.

7. Always be careful to keep clothing and your hands from getting caught in the rotating shaft of the motor. Be ready to turn off power immediately at the first sign of problems. *Read this step completely before applying power.* Install your ammeter around the wire connecting L₁ of the disconnect to L₁ of the motor. As the motor starts, watch for the locked rotor current (LRA), and as the motor runs this current should fall to the full load current listed on the data plates as *amps*. As the motor starts and runs up to full RPM (approximately two seconds) you should hear the click of the end switch. If the current does not return to the rated amperage after the motor is running, turn the power off immediately and call your instructor. Set the ammeter to 30 amp range and apply power at this time.

8. After the motor has run for several minutes, shut off the power and listen for the end switch to click as the shaft rolls to a stop.

9. Restart the motor several times following steps 7 and 8. Be sure you can identify the click of the end switch.

10. Turn the power off and continue.

REVERSING THE ROTATION

At times you may be on a job that requires that the motor shaft run counterclockwise, and the split phase motor you carry as a replacement part runs clockwise. If the motor H.P., speed rating, and voltage are similar you can generally change the shaft rotation.

In the split phase motor the run winding and start winding are identified with a number and color instead of a letter (see in figure 11-5).

In some motors the start and run windings are each made from two windings in series. From figure 11-5 you can see the top run winding is numbered 1 and 2 and the bottom half is numbered 3 and 4. To make one long series circuit, windings 2 and 3 are connected. This makes 1 and 4 the two ends of the run winding. The top start winding is number 5 and 6 and the

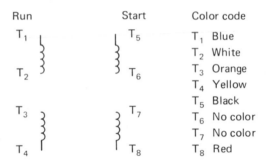

Run	Start	Color code	
T₁	T₅	T₁	Blue
		T₂	White
T₂	T₆	T₃	Orange
		T₄	Yellow
		T₅	Black
T₃	T₇	T₆	No color
		T₇	No color
T₄	T₈	T₈	Red

Figure 11-5   Diagram of split phase motor windings

bottom winding is number 7 and 8. When connected in series, 6 and 7 are connected together, making the start winding terminals 5 and 8. To reverse the rotation of your split phase, simply reverse the terminals of the start winding. In other words, put wire number 8 where number 5 is, and number 5 where number 8 is. On your motor's terminal board these wires will have "spade" or push-on type connectors. Use the following procedures to reverse the shaft rotation on your split phase motor. Check off the steps as you complete them.

1. Turn the power on and start your motor using steps 7 and 8 of the previous procedure. Be sure to stay clear of the shaft. This time check the direction of the shaft's rotation. To determine the direction of rotation, look at the motor from the end opposite to the shaft. Then identify the shaft as turning clockwise or counterclockwise.
   Record the shaft's direction of rotation_____

2. Turn power to the trainer off.

3. Open the inspection cover and locate terminals 5 and 8 on the terminal board.

4. Reverse the position of the wires on terminal 5 and 8. (Put the wire that was on terminal 5 on terminal 8 and the wire that was on terminal 8 on terminal 5.)

5. Call your instructor to check this change.
   Instructor's approval _____

6. Restart the motor and record the direction of rotation _____
   If the direction did not change, call your instructor to inspect your motor.

7. Turn off the power to your trainer and continue with this unit.

CHANGING THE VOLTAGE

Split-Phase Open-Type Motor

114

As noted before, nearly every split phase motor can have its shaft rotation changed. If your split phase motor's data plate has its voltage requirement listed as 115/230, you can change the voltage requirement easily. If the

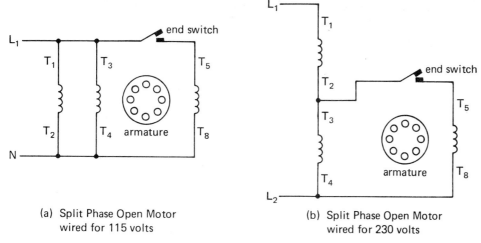

(a) Split Phase Open Motor
wired for 115 volts

(b) Split Phase Open Motor
wired for 230 volts

Figure 11-6   Diagrams of (a) 115 and (b) 230 volt split phase open motor

motor's voltage is listed as one voltage (either 230 or 115) you can not change its voltage requirement.

Remember that the motor must be wired so that its voltage requirement matches the supply voltage of the air conditioner. The diagram in figure 11-6 shows the changes required for 115 and 230 volt motors.

From figure 11-6 you can see that the two parts of the run winding, 1 to 2 and 3 to 4, are parallel with each other for 115 volts and in series for 230 volts. Also note that the start winding (5 to 8) will get only 115 volts even when the supply is 230 volts.

Use the following procedure to change the voltage requirement of your motor. Check off the steps as you complete them. Be sure the motor you use for this activity is rated for 230/115 volts.

1. Make sure the power to your trainer is turned off.

2. Remove all previous wires that connect the disconnect to the motor.

3. Open the inspection plate and find the wires 1, 3, and 5. Connect them together on terminal L_1.

4. Find wires 2, 4, and 8 and connect them together on terminal L_2. This configuration wires your motor for 115 volts. (This is sometimes listed as low voltage.)

5. Connect a wire from the single phase disconnect terminal L_1 to terminal L_1 on the motor.

6. Connect a wire from the neutral bar in the single phase disconnect to terminal L_2 in the motor.

7. Ask your instructor to inspect your wiring.

    Instructor's approval _____

8. Install the ammeter around the wire connected to L_1 and set the meter for 30 amps.

9. Apply power to the motor and follow precautions for starting the split phase motor listed previously in this unit.

10. Record the full load amperage of the motor after it has run for 15 seconds_____.

11. Turn off the power to the trainer.

12. Reconnect the wires in the motor so that wire 1 is connected to L_1 in the motor.

13. Connect wires 2, 3, and 5 together.

14. Connect wires 4 and 8 together and put them on terminal L_2 in the motor. The motor is now wired for 230 volts. (This is sometimes listed as high voltage.)

15. Connect a wire from terminal L_1 of the disconnect to L_1 of the motor.

16. Connect a wire from terminal L_2 of the disconnect to terminal L_2 of the motor.

17. Ask your instructor to inspect your wiring.

    Instructor's approval _____

18. Connect the ammeter around the wire going to terminal L_1. Set range for 30 amps.

19. Apply power and start the motor. Observe precautions as listed previously.

20. Record the full load current after the motor has run for 15 seconds.
    _____.

    Compare this current to the FLA current when the motor was wired for 115 volts. You should notice that the current was doubled when the motor was wired for 115 volts. For this reason some equipment manufacturers use 230 volts, so that that current will be less and they can use smaller size wire to supply power to the air conditioner.

21. Turn the power off and continue with this unit.

TROUBLESHOOTING THE SPLIT-PHASE OPEN MOTOR

The last part of this unit deals with troubleshooting and repair of the split-phase open motor. You should understand that most shops will not attempt any repair on open-type motors under 1 H.P. The reason for this is that customers like to have all work on motors warrantied. For instance, if you put a new end switch in a three year old open-type motor and four weeks later the bearing fails, the customer would expect the motor to be repaired free under warranty. The contractor can't afford to repair the motor again for free. The customer can't afford to put more money into removing the motor and making new repairs.

For this reason most shops replace any motors that they find inoperative. The new motor carries a warranty of from 2 months to 12 months. In this way the customer and the air conditioning repair shop will both be happy.

This is the approach that will be taken in these units. Your job as a technician is to be 100% certain that the motor you determine to be inoperative is actually bad.

Use the following procedure to troubleshoot the split-phase open motor. Check off the steps as you complete them.

1. Make sure the power to your trainer is turned off. Ask your instructor to insert a problem into the motor.

2. Place the ammeter around the wire connected to L_1 and try to make the motor run. If the motor starts and FLA is within limits, the motor is operating correctly. Ask for a new problem if you wish.

3. If the motor draws high current or fails to start, immediately turn off the power at the disconnect.

4. If the motor tried to start and draws high current, skip to step 6.

5. If the motor did not try to start and the current draw was zero, skip to step 10.

6. Since the motor tried to start and was drawing high current, you can assume that voltage has reached the motor.

   a. Try to turn the shaft. If it is difficult to turn you have found a bad bearing. Replace the motor and return to step 2. If the shaft turns freely skip to step b.

   b. Carefully measure the exact voltage at terminals L_1 and L_2 as you try to start the motor. *Be extra careful that your meter probes touch ony terminals L_1 and L_2 and not the side of the motor, or a short circuit will occur.* Observe the voltage as you try to start the motor for two seconds. Shut the power off immediately after two seconds.

   Record the voltage here _____

7. Make sure the voltage at the supply matches thc voltage required on the motor data plate. Make any change as required and repeat this procedure starting with step 2. If voltages match, continue to step 8.

8. Turn off the power and use the diagram in figure 11-6 to be certain all internal motor connections are as listed. If you make any corrections, return to step 2. If no corrections are made, continue to step 9.

9. Try to make the motor run again. This time carefully give the shaft a spin as you apply power. *Be very careful to keep all clothing from getting caught in the shaft.* Be ready to turn off the power immediately at the first sign of any problems.

   a. If the motor tries to start and runs, you have problems with the end switch or start winding. Replace the motor and return to step 2.

b. If the motor still fails to start, it has internal problems and must be replaced. Replace the motor and return to step 2.

10. If the ammeter indicated zero current when you tried to start the motor, skip back only to step 6b to measure the voltage at terminal L_1 and L_2 and return to step 10a.

a. If voltage is present at terminal L_1 and L_2, check the motor overload. If the motor is warm to the touch, the overload is probably open. Allow the motor to cool for 20 minutes and repeat step 2. If current still remains at zero, replace the motor and return to step 2.

b. If no voltage or improper voltage is present at L_1 and L_2, take the proper steps to remedy this problem as outlined in previous units. When proper voltage is supplied, repeat step 2.

When you have completed the troubleshooting sequence, turn off the power and remove all wires from the split phase motor and disconnect. When you are ready, ask your instructor for the quiz on Unit 11.

12 Capacitor-Start, Induction-Run Open Type Motor

Safety for Unit 12

At times the circuits and disconnects you will be working with will be powered with up to 230 volts. You must be aware of and take appropriate safety precautions relative to electrical shock hazards and wear safety glasses while working. A start capacitor will be used during this lesson. You should follow the instructions listed in Unit 8 to discharge and handle the start capacitor. You will also be operating a motor with an open rotating shaft. You must keep loose clothing and your hands clear of this shaft when it is running.

Tools Required

Voltmeter (0-250 volt scale)

Screwdriver (flat blade ¼ in.)

Clamp-on ammeter (30 amp capacity)

Needle-nose pliers (3 in. with insulated handles)

Objectives for Unit 12

At the conclusion of this lesson the student should be able:

To explain to the instructor's satisfaction the theory of operation of the capacitor-start, induction-run open type motor.

Given split phase and capacitor-start, induction-run open type motors, to identify correctly the capacitor-start, induction-run open type motor.

To identify correctly either orally or in writing three uses for the open type capacitor-start, induction-run motor.

To wire correctly the capacitor-start, induction-run motor for a change of voltage and test run the motor on 115 volts and 230 volts.

To wire correctly the capacitor-start, induction-run open type motor for a change of rotation, and test run.

To diagnose accurately electrical and mechanical malfunctions in the capacitor-start, induction-run open type motor and its electrical system.

USES FOR THE MOTOR

The capacitor-start, induction-run (CSIR) open type motor (see figure 12-1) is used in applications where more starting torque is needed. The split phase motor that was discussed in Unit 11 is limited to low-starting-torque applications like fans driven by belts and pulleys, and small pumps on fuel oil furnaces. The CSIR motor is capable of starting heavy water pumps and large furnace fans. In some cases the CSIR open motor is used to drive open-type compressors. These compressors are different from hermetic compressors in that there is no electric motor inside the compressor. Instead the compressor has a shaft protruding through its case with a pulley mounted on it. The CSIR motor turns the compressor shaft with a rubber drive belt. This allows the technician to make a motor change without disturbing the refrigeration unit, as would be required to change the hermetic motor. The CSIR motor driving the open-type compressor is generally found in small commercial refrigeration systems like milk coolers and grocery store display freezers.

Figure 12-1   Capacitor-start, induction-run motor (courtesy of *General Electric*)

THEORY OF OPERATION

The CSIR open-type motor uses the same theory as the CSIR compressor to start and run, except the open type motor uses an end switch to disconnect the start winding at 75% full RPM. The end switch is exactly like the one used

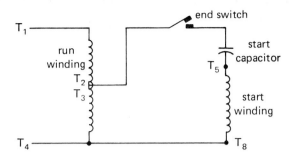

Figure 12-2   Diagram of capacitor-start, induction-run motor

in the split-phase open motor. Figure 12-2 shows a start capacitor connected in series with the start winding.

The capacitor provides a phase shift as in the CSIR compressor. The phase shift is larger than in the split phase motor. This gives the CSIR open motor more starting torque than the split phase motor.

TESTING THE START CAPACITOR

The start capacitor for the CSIR open motor is usually mounted right on top of the motor (in figure 12-1).

Use the following procedure to test the start capacitor on your CSIR motor. Check off the steps as you complete them.

1.   Turn off the power to your trainer.

2.   Locate the CSIR open motor on your trainer.

3.   Remove the two mounting screws securing the steel cover over the start capacitor. Replace the screws after the cover is removed.

4.   If the capacitor does not have a bleed resistor across its terminals, use the procedure in Unit 8 to discharge the capacitor. The capacitor will have one wire connected to each of its terminals. Carefully disconnect the wires from the terminals with a pair of needle-nose pliers. Since the wires can be connected to either capacitor terminal on a start capacitor, you do not need to mark the terminals for reference.

5.   Place the capacitor on your work table and use an ohmmeter to test the capacitor for opens and shorts using the procedure in Unit 8. Answer the following questions about your test.

    a.   Is the capacitor shorted? _____

    b.   Does the capacitor have an open? _____

6.   If the capacitor is defective, replace it with a new start capacitor of the same value. If the capacitor is not shorted or open, replace the two wires to their terminals.

7.   Return the capacitor to the motor and replace the cover. Be sure when you replace the cover that the metal wire terminals do not touch the metal cover.

TEST RUNNING THE CSIR MOTOR

Whenever you replace a motor and test run it, be sure to begin the procedure by taking the data from the data plate. Some technicians feel this is a waste of time—*until* they connect a motor to the wrong voltage or use a motor too small for the job. Remember that in most shops the technicians, not the customer, pays for the mistake (burned out motor).

1. Take the following data from your CSIR motor:

    Model #_____ S.F._____

    Serial # _____ Rotation _____

    Voltage _____   RPM_____

    Amps_____   Frequency _____

    H.P. _____   Phase _____

2. Open the inspection plate and determine if the motor is wired to run for 230 or 115 volts. Remember from Unit 11 that for 115 volts terminals T_1, T_3, and T_5 will be connected to L_1 and T_2, T_4, and T_8 will be connected to L_2. For 230 volts T_1 is connected together and T_4 and T_8 are connected to L_2.

3. Be sure all power to your trainer is turned off.

    a. If your motor is wired for 115 volts connect a wire from load side terminal T_1 (may be labeled L_1) in your motor. Connect a wire from neutral in your single phase disconnect to T_2 (may be labeled L_2) in your motor.

    b. If your motor is wired for 230 volts make the same connection as in step 3a for L_1 but use terminal L_2 in the single phase disconnect instead of neutral.

4. Place your clamp-on ammeter around the wire leading to L_1 and set the scale to 30 amps.

5. Follow the starting instructions and precautions listed in Unit 11 for starting the split phase motor. Be sure you understand the precautions listed in Unit 11 about staying clear of the motor shaft.

6. Start the motor by turning on the power. If the motor does not start or if current is excessive, turn the power off immediately and call your instructor to check your wiring.

7. After the motor has run for a few minutes, record its full load amperage_____.

8. After you have recorded the FLA, turn the motor off and restart it several more times using steps 5 and 6. Listen for the distinct click of the fly weights and centrifugal switch (end switch) as the shaft rolls to a stop and when it is starting.

9. Turn the power off to your motor and wire it for a change of voltage. If it was wired for 230 volts change it to 115 volts. If it was wired for 115

volts change it to 230 volts. Follow the terminal connections described in step 3. Be sure to change your wiring at the load-side terminals of the single phase disconnect so that the supply voltage matches the new requirement for the motor. This means use L_2 for 230 volts and N for 115 volts.

10. Call your instructor to approve your wiring changes.

    Instructor's approval _____

11. Leave the ammeter around the wire connected to L_1. Follow the procedures and precautions for starting the motor. Apply power and observe the ammeter. If the motor does not start or if current becomes excessive, turn the power off immediately and call your instructor to check your wiring.

12. Record the FLA_____.

13. Compare the FLA in step 7 to the FLA in step 12. Notice that the lower voltage causes the higher current and the higher voltage causes the smaller current. This is true for all motors.

14. Turn the power off and leave the wires connected to your motor.

REVERSING THE CSIR MOTOR

The capacitor-start, induction-run motor can be reversed the same way as the split phase motor. All that is required is to change wires T_5 and T_8 around, in other words put wire T_8 in place of T_5 and wire T_5 in place of T_8 (see note in figure 12-3).

    As an air conditioning and refrigeration technician, you may need to reverse a motor's direction of rotation. For example, you might find a bad water pump motor and decide that a change must be made. You find the same size CSIR motor in your truck but its rotation is the wrong direction. If you don't know how to reverse the motor's rotation you will have to drive back to the shop, pick up another motor, and return to the job. All this might take an

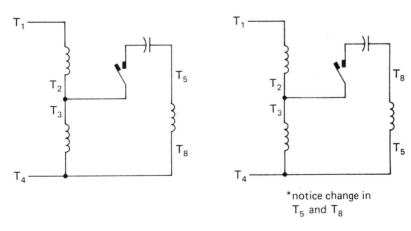

*notice change in T_5 and T_8

Figure 12-3   Diagram for reversing the 230 volt CSIR motor

extra hour and cost the customer an extra $30. A good technician would save the customer extra time and money by simply swapping T_5 and T_8 to reverse the rotation.

Use the following procedure to reverse the direction of rotation of your CSIR motor. Check off the steps as you complete them.

1. Turn your motor on for a few seconds to determine the direction of rotation. Be sure to follow all your starting instructions and observe safety precautions. (Remember, rotation is determined by looking at the motor from the end opposite the shaft.)

   The shaft rotation is_____.

2. Turn the power off.

3. Locate wires T_8 and T_5 and interchange them (put wire T_5 where wire T_8 is and put wire T_8 where T_5 is).

4. Repeat step 1 of this procedure.

   Now the shaft rotation is_____.

If the motor rotation in step 4 is not the opposite direction of step 1, call your instructor to inspect your wiring. Leave the wires to your motor in place and turn off the power to your trainer.

TROUBLESHOOTING THE CSIR OPEN MOTOR

As a service technician you will get calls to check inoperative air conditioning and refrigeration equipment. You must be able to find the problem and make repairs quickly and accurately. The air conditioning and refrigeration business is full of technicians who feel this is a "green light" to replace parts at random until the system starts to run again. Only the trained technician understands that each part of the equipment is easily tested, and that only the inoperative parts need to be replaced. This is why it is imperative that you know the correct method for testing each part. It is also important to know *when* a part should be tested. You will not save time by testing every part of a system.

The next procedure is a fast and accurate method of testing the CSIR open motor. Follow the steps and check them off as you complete them. Ask your instructor to insert a problem into the motor or circuit.

1. Try to make the motor run. As part of this step you should also be able to positively identify the type of motor you are working with. Determine if the motor is open or hermetic. If it is an open-type motor, is it a CSIR or a split phase? This identification is important because it will help you remember the parts that are needed to start the motor you are working on.

   Follow the starting procedures and safety precautions.

   a. If the motor does not try to start and does not make a noise (current draw is zero), skip to step 2.

b. If the motor tries to run and makes a noise but current is excessive, turn the power off and skip to step 3.

c. If the motor runs, measure and record the FLA. If it is within limits you have completed the procedure. If the FLA is too high skip to step 3.

2. Since the motor did not try to start and current draw was zero, the problem is either no voltage or an open in the circuit. (These assumptions can be made in any circuit where current is zero.)

  a. Check for voltage at the load side of the single phase disconnect (L_1 to L_2 for 230 volt motors and L_1 to N for 115 volt motors).

   Record the voltage here_____.

  b. If no voltage is present check the line-side terminals. If you have voltage on the line-side terminals, one or both of the fuses are bad. Replace and return to step 1.

  c. If no voltage is present at the line-side terminals, check the power supply to your room. Reset the circuit breakers and return to step 1.

  d. If proper voltage was present at the load-side terminal, test for voltage at terminals L_1 and L_2 on the motor's terminal board. Be careful that the voltmeter probes do not touch the sides of the motor and cause a short circuit.

   Record the voltage at the motor terminals_____.

  e. If voltage is zero, you have a bad wire between the disconnect and the terminals. Use the voltage drop test you learned in Units 6, 7, and 8 to find the bad wire. Replace the bad wire and return to step 1.

  f. If the current voltage is present on motor terminals L_1 and L_2, put your hand on the motor to see if it is overheated. If so, wait 10 minutes for the overload to cool, close, and return to step 1.

  g. If it is cool to the touch, the motor is bad, no further tests are required, and the motor must be replaced. Remember that changing the motor instead of trying to fix it will be cheaper in the long run and more reliable for the customer.

3. Since your motor tried to run and pulled excessive current, you can assume that the fuse and wiring are in good condition. The problems could be low voltage, bad starting capacitor, wired for wrong voltage, or a problem inside the motor such as a bad end switch that you will *not* attempt to repair. Use the troubleshooting steps that apply to your motor. (Remember: do not allow the power to stay on longer than two seconds for these tests.)

  a. Check to be sure the motor shaft is free to turn. If the motor shaft will not turn easily, replace the motor and return to step 1.

  b. Check for low voltage at the motor terminals as you did in step 2d. If the voltage is not the proper value, use parts a and b of step 2. Make corrections as required and return to step 1.

  c. Check to be sure the motor is wired for the voltage that you are measuring. Make changes if necessary and return to step 1.

d. If the voltage is proper, loosen the cover on the capacitor and place your ammeter around one of the wires going to the capacitor. Try to start the motor again.

e. If there is current in this line, turn off the power and remove the capacitor and test it according to the steps in Unit 8. Replace the capacitor if necessary and try to start the motor. If it does not start, you must assume that it has an internal problem and must be replaced. Remember, you will not attempt to make any repairs inside the motor. Any time you have established the proper amount of voltage at L_1 and L_2 of the motor and determined that the capacitor is good but the motor will not run, you can safely change the motor with the knowledge that you are taking the proper action.

When you have completed the troubleshooting procedure, turn the power off and return the wires to their proper places.

Review the objectives and be sure you can complete each one. When you are ready, ask your instructor for the quiz on Unit 12.

13 Permanent-Split-Capacitor, Open-Type Motor

Safety for Unit 13

At times the circuits and disconnects you will be working with will be powered with up to 230 volts. You must be aware of and take appropriate safety precautions relative to electrical shock hazards and wear safety glasses while working. In addition, appropriate safety precautions should be used when discharging and handling the run capacitor you will be using in this unit. You will also be operating a motor with an open rotating shaft. You must keep loose clothing and your hands clear of this shaft when it is running.

Tools Required

Voltmeter

Screwdriver (flat blade ¼ in.)

Clamp-on ammeter (30 amp capacity)

Needle-nose pliers (3 in. with insulated handles)

Objectives for Unit 13

At the conclusion of this lesson each student should be able:

To explain to the instructor's satisfaction the theory of operation of the permanent-split-capacitor, open-type motor.

To identify correctly either orally or in writing two uses of the permanent-split-capacitor, open-type motor.

To wire correctly the permanent-split-capacitor, open-type motor to run at high, medium, and low speeds.

Given a diagram of the permanent-split-capacitor, open-type motor, to diagnose accurately electrical problems in the motor's electrical system.

USES FOR THE MOTOR

The permanent-split-capacitor (PSC) open motor (see figure 13-1) is the most widely used open-type motor in air conditioners. The PSC motor is commonly used to power blade or propeller type condenser fans and squirrel cage evaporator fans. Its medium starting torque and efficiency while running make it well suited for these applications. This motor is seldom used to drive pumps or belt driven fans.

Figure 13-1 Permanent split capacitor motor (courtesy of *General Electric*)

THEORY OF OPERATION

Like the PSC hermetic compressor motor, the PSC open motor uses only a run capacitor to aid in starting (see figure 13-2). The run capacitor gives enough phase shift to allow moderate starting torque. It also is left in the circuit all the time the motor is running for better running efficiency. This eliminates the need for a centrifugal switch (end switch) which is usually

Permanent-
Split-
Capacitor,
Open-Type
Motor

132

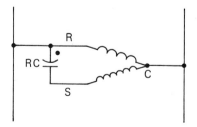

Figure 13-2 Diagram of PSC open motor

the first part to fail in the split phase and capacitor-start, induction-run motors. This means the PSC open-type motor will be more reliable in the field.

TESTING THE RUN CAPACITOR

The run capacitor for the PSC open type motor is usually rated at approximately 3 to 5 uf. It may be mounted on top of the PSC motor or on the electrical panel. Locate the run capacitor for your PSC motor and use the following procedure to test it. Check off the steps as you complete them.

1. Turn off the power to your trainer.

2. Touch a 1K resistor across the terminals to discharge the run capacitor.

3. Set the ohmmeter on R × 10 and zero the meter. Use the procedure in Unit 8 to test the capacitor for opens and shorts. After making the tests, answer the following questions about the condition of your run capacitor.
   a. Is your capacitor shorted? _____
   b. Does your capacitor have an open? _____

4. If the capacitor is defective, replace it with a new one that has exactly the same values in microfarads and volts. It is extremely important that an exactly equivalent capacitor be used when a replacement must be made. If the replacement capacitor is too large (higher microfarad value) or too small (lower microfarad value), it will cause an *increase* in the amount of current in the start winding when the motor is running and cause the start winding to burn out.
   If you needed to replace the capacitor on your trainer, return to step 1 of this procedure and test the new run capacitor for opens and shorts.

TEST RUNNING THE PSC OPEN MOTOR

Before connecting any PSC motor to voltage and test running it, be sure to record the information from the motor's data plate. This step will remind you of the proper voltage, phase, frequency, and H.P. that the motor requires. Remember, a mistake here will usually cause the motor to burn out.
   Check off the steps of this procedure as you complete them.

1. Take the following data from your PSC motor:

   Model # _____ S.F. _____
   Serial # _____ Rotation _____
   Voltage _____ RPM _____
   Amps _____ Frequency _____
   H.P. _____ Phase _____

You may have to remove the PSC motor from its mounting frame to find all the data. To remove the motor from its frame, loosen (do not remove) the two screws from the mounting clamps at each end of the motor (see figure 13-3). Place the mounting clamps where they will not be lost. You will need them to secure the motor in the frame before running the motor.

2. Locate the wires coming out of the stator of the motor. The PSC motor will have three to six wires depending on how many speeds it has. The PSC open motor does not have an inspection plate like the split phase and capacitor-start, induction-run motors. This is the best way to tell these three motors apart. It also means the PSC open motor must be tested like the PSC hermetic motor to identify the wires connected to start, run, and common. Review Unit 6 if you need help to complete the identification.

Figure 13-3   Motor mounting clamps

3. Use your ohmmeter to identify wires connected to start, run and common. Sometimes this color code is used by the motor manufacturers:
   White—Run
   Black—Common
   Brown—Start
   Identify your wires and their colors below:
   Start _____
   Run _____
   Common _____

4. If the motor has more than three wires the motor will have more than one speed. The color code for multi-speed motors is (*Note: some motors may have a different color code*):
   White—Run   Brown—Start
   Black—High Speed   Blue—Medium Speed
   Red—Low Speed   Green—Ground
   It is difficult to identify windings of the multispeed motor with an ohmmeter, therefore you must rely on the manufacturer's color code.

5. Connect the motor wires to the terminals as shown in figure 13-4.
   a. Connect a wire from load-side terminal L_1 in the single phase disconnect to the "red dot" terminal of the run capacitor.

Permanent-
Split-
Capacitor,
Open-Type
Motor

134

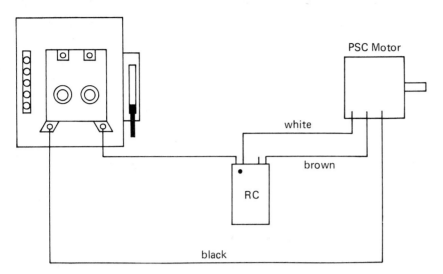

Figure 13-4   Diagram of PSC motor wired to disconnect

   b. Connect the white (run) motor wire to the "red dot" terminal of the
   run capacitor.

   c. Connect the brown (start) motor wire to the other capacitor terminal.

   d. Connect the black (common) motor wire to the L_2 load-side terminal
   in the single phase disconnect.

   e. Put wire nuts on all other unused wires.

6. Have your instructor check the wiring you have completed in step 5.

   Instructor's approval _____

7. Place the clamp-on ammeter around the brown (start) wire and set the
range to 15 amps.

8. *Before applying power read this completely.* The start winding current
for the PSC motor should rise to approximately 6 to 10 amps and then
return to 3 to 5 amps during run. If the motor fails to start or the start
winding current remains above the 3 to 5 amp range after the motor is
running, turn the power off *immediately* and call your instructor to
inspect your motor. Also be sure to observe the safety precautions
concerning the open rotating shaft listed in Unit 11.

   Apply power at this time and record the amperage in the start
winding when the motor is running.

   _____ amps.

9. Turn the power off to your motor. You should notice there is *no* click
when the PSC motor stops and starts. This is because there is no end
switch in the PSC motor. Restart the motor several more times using
step 8. You do not need to wait to restart open motors since there is no
refrigerant pressure to balance as in the hermetic motors.

   When you have completed step 9, leave the wires in place and turn
off the power to the trainer.

WIRING THE PSC MOTOR FOR A CHANGE OF VOLTAGE AND A CHANGE OF ROTATION

Because the wires come out of the side of the stator and there is no inspection plate available, the PSC open-type motor cannot have its wires changed for a different rotation or a different voltage. This means that motors of different voltages for each rotation must be stocked by the equipment dealer. Since the PSC open motor is generally 1/3 or 1/4 horsepower, the total number of motors needed to be stocked is usually eight. Dealers are generally able to afford this arrangement.

There will be no activity to change voltage or rotation on the PSC open type motor. Instead, you must be aware of the importance of the data you recorded in step 1 of the test run procedure. If you know the information about the motor, you should be able to replace it with another PSC motor with exact specifications.

WIRING THE PSC MOTOR FOR A CHANGE OF SPEED

Use the following procedure and figure 13-5 to wire the PSC open motor for a change of speeds (RPM). Check off the steps as you complete them.

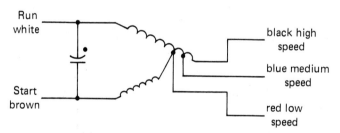

Figure 13–5   Three-speed PSC motor diagram

NOTE: Some manufacturers call the *white* wire "common." This is because the white (run) wire is common to all three speeds. Remember that the black, blue, or red wire is the common terminal for wiring purposes.

1. Turn on the power to your motor. Observe the starting and safety procedures. Allow the motor to run for several minutes and observe the speed. This is the high speed.

2. Place the clamp-on ammeter around the wire connnected between the run capacitor and terminal L_1 in the disconnect and record the FLA
    _____.

3. Shut off the power to the trainer.

4. Remove the black motor wire from L_2 of the disconnect and put a wire nut on it.

*Permanent-
Split-
Capacitor,
Open-Type
Motor*

136

5. Remove the wire nut from the (blue) motor wire and connect it to terminal L_2 on the load side of the single phase disconnect. This connects the motor for medium speed.

6. Instructor's approval_____.

7. Start the PSC motor and observe all starting and safety precautions.
   Record the FLA for this speed_____.
   You should notice the speed now is slower than the speed in step 1. The FLA should be less.

8. Turn the power off to the trainer.

9. Remove the blue wire from terminal L_2 of the disconnect and replace the wire nut on it.

10. Remove the wire nut from the red wire and connect it to terminal L_2 in the single phase disconnect. This connects the motor for the low speed.

11. Start the motor wired for low speed while observing the starting and safety precautions.
    Record the FLA for low speed_____.
    Compare this speed to the motor's speed in step 1 and step 7. You should notice the motor speed is slower and the FLA is lower.

12. Turn the power to your trainer off and leave all wiring in place.

TROUBLESHOOTING THE PSC OPEN MOTOR

Troubleshooting the PSC motor is easily accomplished in the field by the air conditioning and refrigeration technician. Just as with the split phase and CSIR open motors, repairs to the motor are not made by the technician. As explained in Units 11 and 12, the open type motor is easily diagnosed and it is quicker and more reliable to replace the motor than repair it. Therefore, it is most important that the technician be positive a motor is bad before he or she replaces it. Use the following procedure to diagnose electrical problems in the PSC open-type motor. Check off the steps as you complete them. Ask your instructor to insert a problem into your PSC motor or circuit at this time.

1. Try to start the motor. Place the clamp-on ammeter around the wire connecting L_1 to the run capacitor. Observe all safety and starting precautions as you apply power and try to start the motor.

   a. If the motor hums or tries to start and draws excessive locked rotor amperage (LRA), shut the power off immediately and skip to step 2.

   b. If the motor does not start and does not make any noise (current draw is zero), leave the power on and skip to step 3.

   c. If the motor runs and the current level is acceptable (within data plate specifications), shut the power off. The motor is operating correctly. Ask your instructor for a new problem if you need more practice. If you have completed the troubleshooting, turn the power off to your trainer and replace all wires and parts.

2. Since the motor tried to start and used excessive current, you must assume that you have applied voltage and a complete circuit through part of the motor. Check for the following problems, make repairs as needed, and return to step 1.

   a. Stuck motor shaft (bad bearing)

   b. Bad run capacitor (use ohm test)

   c. Bad start winding (use ohm test)

3. Since the motor did not try to run and there was no current, make the following checks, make repairs as needed, and return to step 1.

   a. Check for applied voltage at the load-side terminals of the single phase disconnect. If voltage is zero, replace fuses and return to step 1. If voltage is full applied, skip to step 3b.

   b. Check for applied voltage at the motor wire terminals of the run winding (white wire) and the common (black wire for high speed). If voltage is zero find the bad connection or wire, make appropriate repairs, and return to step 1. If voltage is full applied, check the motor for an open winding. Replace the motor and return to step 1.

This completes the troubleshooting activity for Unit 13. Review the objectives for Unit 13. When you are ready, ask for the quiz.

Permanent-
Split-
Capacitor,
Open-Type
Motor

14 Shaded Pole Open-Type Motor

Safety for Unit 14

At times the circuits and disconnects you will be working with will be powered with up to 230 volts. You must be aware of and take appropriate safety precautions relative to electrical shock hazards and wear safety glasses while working. You will also be operating a motor with an open rotating shaft. You must keep loose clothing and your hands clear of this shaft when it is running.

Tools Required

Voltmeter (0-250 volt scale)

Screwdriver (flat blade ¼ in.)

Clamp-on ammeter (30 amp capacity)

Objectives for Unit 14

At the conclusion of this lesson the student should be able:

To explain either orally or in writing the theory of operation of the shaded pole motor.

To draw an accurate electrical diagram of the shaded pole motor.

When given a split phase and shaded pole open-type motor, to identify accurately the shaded pole motor.

To wire the shaded pole motor correctly and test run.

To find a given electrical problem in the electrical circuit of the shaded pole motor to the instructor's satisfaction.

USES FOR THE SHADED POLE OPEN-TYPE MOTOR

The shaded pole open-type motor (see figure 14-1) is a low-starting-torque motor used on small evaporator and condensing fans. Manufacturers of small air conditioners and residential refrigerators use the shaded pole motor because it is not expensive to make and it does not need any starting devices or relays. Because of its low starting torque, the shaded pole motor cannot be used to turn pumps or large fans.

THEORY OF OPERATION

The shaded pole motor is an induction-type motor that has a shading pole (copper bar) added to each field pole in its stator (see figure 14-2). This shading pole takes the place of a start winding by causing a rotating magnetic field. The rotating magnetic field works in a manner similar to the rotating field caused by the start winding in the split phase motor. This helps the motor to start.

The shading pole in the shaded pole motor cannot produce as much torque as a start winding, but it has one advantage. By not having a start winding, all the wire in the motor can be heavier, as in the run winding of a

Figure 14-1   Shaded pole motor (courtesy of *General Electric*)

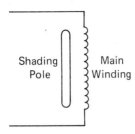

Figure 14–2   Diagram of shaded pole motor

split-phase motor. This means that if the shaded pole motor did not start and stall it would not pull large starting current (LRA), and the motor windings would be able to withstand a prolonged locked rotor current without burning up the motor. Also, since it does not have a start winding, it does not need a starting relay or capacitor.

TESTING THE SHADED POLE MOTOR

Use the following procedure to test the shaded pole motor on your trainer. Remember that this procedure, like all procedures in these units, can also be used when you are servicing and repairing equipment in the field. Check off the steps as you complete them.

1. Turn off all power to your trainer.

2. Locate the shaded pole motor on your trainer.

3. Loosen the mounting screws and remove the shaded pole motor from your trainer.

4. Set your ohmmeter on R × 1 and put one probe on each motor wire. Record the resistance _____

5. Since the motor windings in a shaded pole are all wired in series, the measurement in step 4 will determine if there is a break in any wire.

    a. If the resistance measured in step 4 was infinite (∞) repeat the measurement with the meter on the highest range, R × 100 K, and record _____.
    If the measurement shows there was some resistance in the winding, skip to step 6.

    b. If the measurement in step 5a is still infinite (∞) your motor has an open in its winding. Call your instructor for a new motor and repeat this procedure.

6. Record the data from the data plate on your shaded pole motor.

Model _____	S.F. _____
Serial No._____	Rotation _____
Voltage_____	RPM _____
Amps _____	Frequency _____
H.P._____	Phase _____

7. Return your motor to its mounting bracket and secure it for running.

8. Connect your shaded pole motor to correct voltage to test run. Since your motor has only two leads, you may connect either lead to L_1 on the load side of the single phase disconnect. Connect the other lead to L_2 on the load side also. (If the motor is 115 volts use neutral instead of L_2.)

9. Connect your clamp-on ammeter around either of the lines and set the range for 15 amps.

10. Apply power to your motor and observe all safety and starting precautions that you learned in previous units. If the motor fails to start or draws excessive current, turn the power off immediately and call your instructor.

11. Record the running amperage (FLA) of your motor_____

12. Turn the power off and call your instructor at this time. Ask your instructor to recheck the motor to be sure it is securely mounted before proceeding to the next step.

13. During this step the instructor will hold the shaft of your shaded pole motor to cause it to stall when it tries to start so you can record locked rotor amps (LRA) *Be sure to use a good quality leather glove when holding the shaft.* This motor is a very low torque motor and can easily be stalled without harming the motor. *Do not try this procedure with any other motor.*

    a. At this time have your instructor hold the shaft for two to three seconds as he starts the motor and then release the shaft and allows the motor to run.
    Record the LRA _____
    You should notice that LRA for the shaded pole motor is not significantly higher than the running amperage FLA. In the other hermetic and open type motors the LRA is three to fives times larger than the FLA.

14. Turn the power off to the trainer and leave all wiring in place.

REVERSING THE SHADED POLE MOTOR

You may encounter a refrigeration unit or air conditioner that requires the new shaded pole motor you are installing to rotate in the opposite direction. This is accomplished in the shaded pole motor by carefully removing the rotor and reversing it. Use the following procedure to reverse the rotation of your shaded pole motor. Check off the steps as you complete them.

1. Turn the power on and run your shaded pole motor long enough to determine its direction of rotation. Remember to determine the direction of rotation of all open motors by observing the shaft from the end opposite the shaft (inspection plate end). The motor rotation is

2.   Turn the power off and remove the two motor wires from the disconnect.

3.   Remove the motor from its mounting bracket.

4.   Mark the end plates of the motor as instructed in Unit 6, so that you will be able to return them to their exact locations during reassembly.

5.   Loosen and remove the four nuts and bolts that hold the end plates in place. Keep these parts where they will not be lost. You will need them for reassembly.

6.   Carefully tap the end plates with a screwdriver handle to loosen them.

7.   Before removing the rotor from the stator, use a pencil or mark an arrow on the stator indicating the end the shaft was protruding from the stator. Remove the end plates and be careful not to disturb the bearings. Be sure to leave any washers on the rotor shaft in place.

8.   Carefully remove the rotor and lay it in a *clean* place where it can not fall on the floor.

9.   Now that the rotor is removed from the stator, look into the stator and notice the size of the shading pole (the copper bar placed in the slot of each winding). Figure 14-3 shows the location of the shading pole.

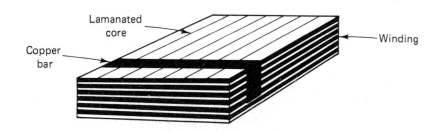

Figure 14–3   Diagram of shading pole

10. Carefully replace the rotor into the stator so the shaft points in the direction opposite the arrow you marked in step 7.

11. Carefully replace the end plates. Be sure to align the marks so the end plates are returned to their original location and the oil holes are pointing up.

12. Replace the four bolts and nuts that hold the end plates in place and tighten them.

13. Remount the motor in its bracket.

14. Rotate the shaft by hand. If it does not rotate freely, call your instructor to inspect your reassembly. Be sure the shaft is rotating freely before continuing to step 15.

15. Follow step 8 of the testing procedure, and reconnect the motor wires to the single phase disconnect.

16. Follow the safety and starting precautions and restart the motor by applying power.

Record the FLA_____.

17. Observe the direction of rotation. Notice the direction is opposite what it was in step 1. (If the motor shaft is turning the same direction as step 1, call your instructor to inspect your motor.)

18. Turn off the power to your motor and leave all wires in place for the troubleshooting procedure.

TROUBLESHOOTING THE SHADED POLE MOTOR

At times you will be called to repair an air conditioner or refrigerator that has an inoperative shaded pole motor. You must be able to identify the motor as a shaded pole motor. Remember that it has no capacitors and that there is no inspection plate since there is no end switch. If it is single speed it will have only two wires for supplying power.

Use the following procedure to diagnose the problems in your shaded pole motor. Check off the steps as you complete them. Ask your instructor to insert a problem at this time.

1. Try to start the motor. Observe all safety and starting precautions.

    a. If the motor tries to start but draws excessive current, turn the power off immediately and skip to step 2.

    b. If the motor does not try to start and does not draw any current, skip to step 3.

    c. If the motor starts and full load current is within specifications, turn the power off and ask for a new problem if you need more practice.

2. Since the motor was using current, you can make two assumptions: (1) there is voltage to the motor, (2) the winding is operating correctly. The problem is in the shading pole or bearing.

    a. Check the shaft for freedom of movement. If the shaft does not turn freely, replace the motor and return to step 1.

    b. If the shaft turns freely you can assume the shading pole is damaged. Replace the motor and return to step 1.

3. If the motor does not start and does not draw any current, make the following checks:

    a. Check for proper voltage at terminals L_1 and L_2. Replace fuses and restore power if necessary. Return to step 1.

    b. If proper voltage level is available at L_1 and L_2, turn off power and check the motor winding using the procedure listed in this unit. Replace the motor if necessary and return to step 1.

This completes all activities for Unit 14. Call your instructor when you are ready for the quiz on Unit 14.

15 Three Phase Motors

Tools Required

Voltmeter (0-250 volt scale)

Screwdriver (Flat blade ¼ in.)

Clamp-on ammeter (30 amp capacity)

Needle-nose pliers (3 in. with insulated handles)

Objectives for Unit 15

At the conclusion of this lesson the student should be able:

To explain to the instructor's satisfaction the theory of operation of the three phase motor.

To explain to the instructor's satisfaction what will happen to the three phase motor if it loses one of its lines or phases of voltage.

To test correctly each winding of the three phase motor with an ohmmeter.

To wire correctly the three phase motor for low voltage (230 volts) operation and test run the motor. .

To change the wiring correctly so the three phase motor will reverse its rotation.

THREE-PHASE HERMETIC AND OPEN-TYPE INDUCTION MOTORS

On commercial installations where three-phase voltage is available, hermetic and open-type motors will usually be three-phase. Three-phase motors, also called poly-phase motors, provide high starting torque and excellent running efficiency. These motors are used to drive compressors, pumps, and fans in air conditioning and refrigeration systems. The air conditioning and refrigeration technician must be able to wire the three-phase open motors for a change of rotation, a change of voltage, or both, and also be able to troubleshoot and install both hermetic and open-type three-phase motors.

THEORY OF OPERATION

The basic theory of operation is the same for both the hermetic and the open-type three-phase induction motor. The major difference is that the hermetic motors generally have only three terminals (see figure 15-1). These terminals are called T_1, T_2, and T_3. Open-type motors will usually bring the end of each winding out to a terminal board or inspection plate (see figure 15-2).

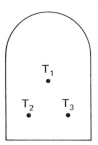

Figure 15–1 Three-phase compressor

Therefore most three-phase open motors have 9 or 12 terminals. These are identified as T_1, T_2, etc. The three-phase motor has three separate windings that are powered by three separate phases of voltage. Since each phase of the three-phase voltage is separated by 120° and the three windings are

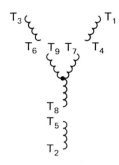

Figure 15-2 Three-phase open motor

displaced in the three phase motor, no starting device or start winding is required to get the magnetic field to rotate in the stator. Since the rotor is a "squirrel cage" rotor, a magnetic field can be induced into it just as in the single-phase induction motors. This means that as soon as three-phase power is provided to the windings the rotor begins to move.

The three windings of the three-phase motor are equal in magnetic strength. This gives the motor good efficiency and makes it more durable. These windings can be wired in a star (wye) configuration or delta (Δ) configuration. The star and delta wired motors are similar to one another in operational performance. The major difference is that the delta wired motor draws less starting current and provides a little less starting torque.

TESTING THE THREE-PHASE OPEN MOTOR

Locate the three-phase open motor on your trainer. Use the following procedure to test the windings of your three-phase open motor. Check off the steps as you complete them.

1. Turn off the power to the trainer.

2. Remove the inspection cover from your motor.

3. Locate the terminals in your motor. Notice that each terminal is numbered. Use the ohmmeter to check each winding.

    a. The winding will be found between the following terminals if the motor is wired for delta:

        T_1 to T_4_____

        T_1 to T_9_____

        T_2 to T_5_____

        T_2 to T_7_____

        T_3 to T_6_____

        T_3 to T_8_____

    b. The windings will be found between the following terminals if the motor is wired for wye (star):

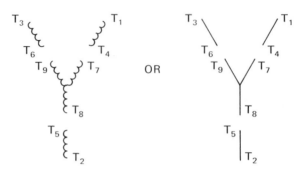

Record the resistance:

T_1 to T_4 _____ T_7 to T_9 _____

T_2 to T_5 _____ T_8 to T_9 _____

T_3 to T_6 _____ T_9 to T_7 _____

The motor on your trainer can be connected as a wye motor. Use the following procedure to test run your three-phase motor in wye configuration. Check off the steps as you complete them.

1. Turn the power off to the trainer.

2. Remove the inspection plate cover from the end of your motor.

3. Remove all terminal connections so that the ends of each wire are free. These ends will be identified by numbers T_1, T_2, etc.

4. Set your ohmmeter on R $\times$ 10 and zero.

5. Measure the resistance through the following terminals and record.

T_1 to T_4 _____ T_7 to T_8 _____

T_2 to T_5 _____ T_8 to T_9 _____

T_3 to T_6 _____ T_9 to T_7 _____

$$T_1 \text{—} \text{mmm} \text{—} T_4$$

$$T_2 \text{—} \text{mmm} \text{—} T_5 \qquad T_7 \text{—} \text{mmm} \begin{cases} \text{—} T_8 \\ \text{—} T_9 \end{cases}$$

$$T_3 \text{—} \text{mmm} \text{—} T_6$$

You can see from the diagram that these are the windings of the motor and that an open (infinity) or a short (zero) resistance will not allow the motor to operate correctly. If any of your readings were zero or infinity, call your instructor and request a new motor.

6. Connect the motor for 230 volts to test run. To do this you will connect the wye windings in parallel (see figure 15-3).

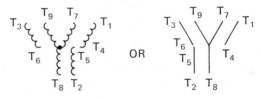

Figure 15-3

To wire the windings in parallel, connect T_1 to T_7, T_2 to T_8, T_3 to T_9, and connect T_4, T_5, and T_6 together.

7. Connect a wire from load-side terminal L_1 of the three-phase disconnect to terminal T_1, T_7.

8. Connect a wire from load-side terminal L_2 of the three-phase disconnect to terminals T_2, T_8.

9. Connect a wire from load-side terminal L_3 to T_3, T_9.

10. Ask your instructor to inspect your wiring.

    Instructor's approval _____

11. Place the clamp-on ammeter around any one of the wires from the disconnect and set the scale to 30 amps. (Note that the ammeter will only read the amperage of one line at a time.)

12. *Read this step completely before applying power.* The starting current (LRA) for the three-phase motor should be the same on each of the three lines. The LRA will drop to the run current level (FLA) approximately two seconds after the motor starts. The FLA should also be equal on all three lines. When you apply power to start your motor, monitor the ammeter. If the motor does not start or if current becomes excessive, turn the power off immediately and call your instructor. Be sure your motor is securely mounted on your trainer. Keep all loose clothing and your hands away from the shaft while the motor is running. Start your motor by applying power at this time.

13. Measure and record the current on all three power lines.

    Line 1 amps _____

    Line 2 amps _____

    Line 3 amps _____

14. Turn the power off and leave all wires in place.

15. The high voltage required to operate your three-phase motor will be approximately 480 volts. Since this voltage is not available, you will not be able to test run the motor on high voltage. (Your instructor may ask you to change the wiring for high voltage but not apply any power.) If you change the motor's wiring to a high voltage mode, repeat step 6 of this procedure to return the motor to low (230 volts) voltage operation. *Be sure your motor is wired for 230 volts before you continue to the next section.*

REVERSING THE ROTATION OF THE THREE-PHASE MOTOR

The three-phase motor can have its rotation reversed by reversing any two power lines. Take the wire connected to L_1 and connect it to L_2, and take the wire that was on L_2 and connect it to L_1. (The same result could have been obtained using L_2 and L_3 or L_1 and L_3.) Use the following procedure to reverse the rotation of the three-phase motor. Check off the steps as you complete them.

1. Restart the three-phase motor, observing the starting and safety precautions. Observe the direction the shaft is rotating and record____

   _____

2. Turn off the power to your trainer and reverse the location of the wires on L_1 and L_2.

3. Restart the motor, observing the starting and safety precautions. Observe the direction in which the shaft is rotating and record_____

   _____

   If the shaft is not turning in a direction opposite to that listed in step 1, call your instructor.

4. Turn the power off and interchange leads L_2 and L_3.

5. Restart the motor and observe the starting and safety precautions. Observe the direction the shaft is rotating and record_____

   _____

   This time the shaft should rotate in a direction *opposite* to that listed in step 3. If the shaft is turning in the *same* direction, call your instructor. If the shaft has reversed direction of rotation, turn the power off and leave all wires in place. Remember, this method of changing rotation of the shaft by reversing any two leads will work for any three-phase motor, including hermetic compressors.

DIAGNOSING THE THREE-PHASE MOTOR FOR ELECTRICAL PROBLEMS

The air conditioning and refrigeration technician must be able to find electrical problems in three-phase motors accurately and quickly. Three-phase motors over 1 H.P. are usually repaired at a motor repair facility. The technician must be able to determine that the motor is actually inoperative before removing it. Use the following procedure to test your three-phase motor for electrical problems. Check off the steps as you complete them. Ask your instructor to insert a problem in your motor at this time.

1. Try to make the motor run. Keep the ammeter in place and observe all safety and starting precautions.

   a. If the motor does not try to start and does not draw any current, skip to step 2.

   b. If the motor tries to start and current draw is excessive, turn the power off immediately and skip to step 3.

   c. If the motor starts and runs and FLA is within limits, shut off the motor. Ask for another problem if you need more practice. If you have completed the troubleshooting procedure, turn off all power and return all wires to their storage locations.

2. Since the motor did not pull any current, the most likely cause is a loss of three-phase voltage. Check for three-phase voltage. Replace fuses as needed and return to step 1.

3. Since the motor tried to start and did use current, the most logical problem is the loss of only one line or phase of voltage. Remember that a three-phase motor *must* have all three lines or phases of voltage present at the motor terminals.

    a. Check for all three-phases of voltage at the disconnect and make repairs as needed. Return to step 1 when voltage is restored.

    b. If voltage is present at the disconnect, check for voltage at the motor. This test could also be accomplished by clamping an ammeter around each line while you allow the motor to try to start for two seconds. The line that has no current draw has lost voltage either in the wire or inside the motor on that phase. Make appropriate repairs and return to step 1.

    c. If you have voltage at all three phases at the motor terminals and the motor will not run, the windings are open. Check the windings for proper terminal connections and proper resistance. Replace the motor if needed and return to step 1.

    This troubleshooting technique is also usable on three-phase hermetic motors. The three-phase hermetic compressor has three terminals, marked T_1, T_2, and T_3. If there is voltage present at each of these three terminals and the compressor will not run, make the resistance test. All three measurements should be the same. The ammeter can also be used. Remember that all three lines should draw the *same* amount of current if the motor is operating correctly.

    This completes the activities for Unit 15. Ask your instructor for the quiz on Unit 15 when you are ready.

16 Relays, Contactors, and Solenoids

Tools Required

VOM (250 volt scale and Rx100K ohm scale)
Screwdriver (¼ in. flat blade)

Objectives for Unit 16

At the conclusion of this unit the student should be able:

To explain orally or in writing the operation of a relay to the instructor's satisfaction.

To explain orally or in writing the operation of a solenoid valve to the instructor's satisfaction.

To explain orally or in writing the difference between a relay and a contactor to the instructor's satisfaction.

To explain orally or in writing the operation of a relay or a solenoid coil that is connected to the improper voltage to the instructor's satisfaction.

To draw a relay correctly in the proper voltage circuit of a ladder diagram.

To check a relay with an ohmmeter and identify correctly the normally open and the normally closed contacts.

To identify correctly the normally open and the normally closed contacts on a relay electrical symbol.

REVIEW OF BASIC RELAY OPERATION

The relay was introduced in Unit 5. It is a magnetically controlled switch that is widely used in air conditioning and refrigeration systems to control the compressor and fan motor.

The relay has two basic parts, the coil and the contacts. These parts are mounted on the same base but are not electrically connected. The coil, made of many turns of wire, becomes a very strong magnet when current is passed through it. When current is stopped, the coil loses its magnetic field. The normally open contacts act as a simple switch that is pulled closed when the coil becomes a magnet (normally closed contacts will pull open when the coil is magnetized). When the coil loses its magnetic field the contacts will return to their normally open position by a spring. This means the coil must be energized to become a magnet before the normally open contacts will close and allow large currents to flow to the compressor and condensor fan. Review Unit 5 if you need more operational information.

IDENTIFICATION OF COIL VOLTAGES

The voltage to energize the coil of the relay found in air conditioning and refrigeration circuits is 24 volts, 120 volts, or 208/230 volts. For this reason there are three coils available. Air conditioning and refrigeration manufacturers can use any of the voltages listed above to power the control circuit for their equipment. The coil and the coil voltage must match exactly. This will be demonstrated by the following activity. Use the procedure below to show that the voltage needed to energize the coil must match the voltage requirement listed on the coil of the relay. Check off the steps as you complete them.

1. Turn off the power to your trainer.

2. Locate all four relays on your trainer.

3. Draw a diagram of their relative locations on the trainer in the space below and label the relays in your diagram A, B, C, and D.

*Relays,
Contactors,
and Solenoids*

158

4. Check the side of the coil on each relay and indicate the required voltage of each on your diagram.

5. With the power turned off, connect the 230 volt coil wires to the L_1 and L_2 terminals of the single phase disconnect. This will supply 230 volts to the 230 volt relay coil.

    a. Apply power to the relay by turning on the disconnect. The contacts should snap closed.

    b. Set your voltmeter to 250 scale and measure and record the voltage going to the coil_____.

    c. Notice that the magnetic field caused by the 230 volts will pull the relay contacts closed with a sharp snap. This confirms the theory that the voltage supply to a relay coil must match the requirement listed on the coil. Another very important point is that you can easily measure the voltage across the coil terminals. If no voltage or incorrect voltage is present, the coil will not pull in the contacts.

6. Locate the relay with the 208/230 volt coil.

    a. Connect the coil of the 208/230 volt relay to the 24 volt terminals R and C of the transformer on your trainer. (Wire your control transformer as per instructions in Unit 4 if it is not wired to the single phase disconnect at this time.)

    b. Apply power to the coil of the 208/230 volt relay by turning on the single phase disconnect. (Allow power to stay on for only 10 to 15 seconds as you complete step c.)

    c. Set your voltmeter to 50 volts, measure the voltage across the coil terminals, and record the voltage going to the coil_____ _____.

    d. Notice that approximately 24 volts is reaching the relay coil. But, since the coil required 208/230 volts, the magnetic field created by the coil is not strong enough to pull the contacts closed.

    e. Repeat steps b and c if you need to see this again. Turn the power off and remove the two coil wires from the 24 volt power source.

7. Connect the 208/230 volt coil to the 120 volt source in the single phase disconnect. Use L_1 and neutral terminals.

    a. Apply 120 volts to the 208/230 relay coil for 10 to 15 seconds while you measure the voltage at the coil terminals. Turn the power off immediately after measuring the voltage.

    b. Record the voltage going to the coil_____.

    c. Notice that 120 volts will not create a strong enough magnetic field in the 208/230 volt coil to pull in the contacts. When the relay coil does not receive enough voltage to pull in the contacts, the relay will hum loudly.

8. Repeat this experiment with the 120 volt relay coil by connecting it to the 24 volt power source and the 120 volt power source. Notice that when the 24 volts is applied to the coil it will hum but not have a strong enough magnetic field to close the relay contacts. The 120 volt relay coil will only operate on 120 volts.
    NOTICE!! Always apply voltage of the correct value as listed on the coil data. Voltage of greater or lesser value will burn out the coil.

From the previous activity you can see that it is important to identify the amount of voltage the coil requires and make sure it receives exactly the amount of voltage it requires.

CHANGING RELAY COILS

Some relays can have their coils changed. This can be accomplished if the coil is bad or if the coil does not match the voltage available. Use the following procedure to change the coil in one of your relays. Check off the steps as you complete them.

1. Turn off the power to the trainer.
2. Unscrew the mounting screws at the coil.
3. Remove the retaining plate and carefully pull the coil off the relay.
4. Replace the coil with a new coil of a different voltage.
5. Replace the retaining plate and mounting screws.
6. Test the relay coil at its new voltage as indicated in steps 1 to 8 above.
7. Turn off power and return all wires to their place.

Some contractors prefer to buy one relay and have a 24, 120 and 208/230 volt coil on hand. Since the coils only cost a few dollars, the relay coil can be replaced or changed to match the existing voltage.

CONTACTS

The contacts for a relay may be symbolized in any of the following ways.

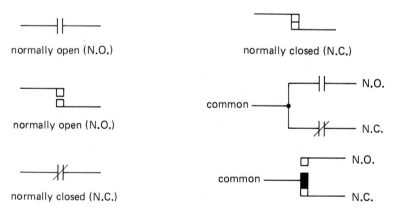

The contacts will be rated in the following areas: maximum allowable voltage, maximum allowable current, and maximum combined voltage and current (VA).

If the relay's contacts are rated to carry more than 15 amps it will be called a *contactor*. This means the contactor operates exactly like a relay except its contacts can carry *more* than 15 amps.

It is important to be able to identify the terminals on a relay or contactor that are connected to the contacts. It is also important to know if the contacts are normally open (when no electricity is applied to the coil) or normally closed. The ohmmeter on R X 1 scale is the best meter to use to check the relay's contacts.

Use the following procedure to locate the terminals connected to the relays contacts and identify them as N.O. or N.C. Check off the steps as you complete them.

1. Turn off power to your trainer.

2. Locate the four relays on your trainer and draw a diagram of their relative locations in the space provided below. Label the relays according to their coil ratings.

3. Record the voltage and current rating of each relay's contacts on the diagram. Identify the relays that have a contact rating of over 15 amps as contactors.

4. Use your ohmmeter to test the terminals on your relays and contactors to determine which terminals are connected to contacts. To complete this step keep the power off and use a screwdriver to open and close the contacts. Add the relative locations of the sets of contacts to your diagram.

   a. If the meter reading is infinity (∞) and moves to zero when the contacts are moved, the terminals you are testing are connected to a set of normally open contacts.

   b. If the meter reading is zero and moves to infinity (∞) when the contacts are moved with the screwdriver, the terminals you are testing are connected to a set of normally closed contacts.

   c. If the meter reading is infinity (∞) before and after you move the contacts, the terminals you are testing are not a contact set. These two terminals will never make a complete circuit.

5. Sometimes on open contact relays it is possible to see which terminals are connected when the contacts close. In the diagram below (figure 16-1) the contacts can easily be seen and the terminals that are connected can be identified.

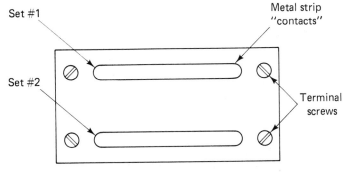

Figure 16-1   Relay contacts

When the contacts close the strip of metal touches a contact point on each end as shown in figure 16-2.

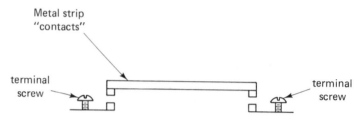

Figure 16-2   Contacts

DIAGRAMMING THE RELAY

The ladder diagram of the relay is generally confusing to beginning technicians because the contacts are usually in the top part of the diagram near the motor and the coil is at the bottom near the thermostat. This fact will not confuse you if you remember that the ladder diagram is only trying to show *sequence of operation*. This means that the voltages in the circuit diagram must be separated. The high voltage (208/230) will usually be at the top. The low voltage circuit will usually be at the bottom. Since the coil on the relays in previous units was usually 24 volts, it was located in the 24 volt circuit at the bottom of the diagram. The contacts usually controlled a 230 volt motor and were in the top part of the diagram. Even though they do not appear to be close together on the diagram, you must remember that the coil is directly under the contacts in real life. They are in separate circuits in the diagram simply because they operate at different voltages.

The following diagrams are an example of this. In the first diagram you will notice the 120 volt compressor is controlled by a relay with a 24 volt relay coil (see figure 16-3).

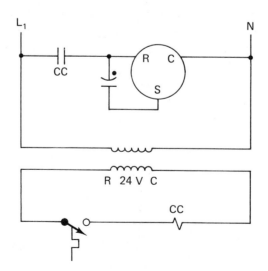

Figure 16-3

Notice in figure 16-3 that the relay will control a rather large amperage to run the compressor, so it is identified as a *contactor*. Both the coil and the contacts have the same identifying compressor contactors or CC, so called because the job of this contactor is to control the compressor.

In the second diagram (figure 16-4) the relay now has a 120 volt coil. It is still labeled CC and still controls the compressor.

Notice that since the coil is now 120 volts, it is in the same circuit as the contacts and compressor, and since there are no components that need 24 volts, the transformer is not required.

In the space below draw the diagram again. This time the compressor requires 230 volts and needs two sets of N.O. contacts, one set for L_1 and one set for L_2. The coil for this relay is 230 volts. Draw the coil in the right voltage circuit. Ask your instructor to check your diagram. Instructor's approval_____

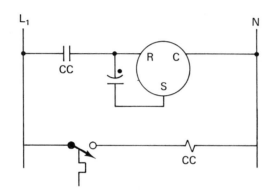

Figure 16–4

SOLENOID VALVES

The solenoid valve (figure 16-5) is used to operate a valve. It has a coil similar to the relay that becomes a very strong magnet when energized. The magnetic coil moves a valve open or closed in a solenoid valve instead of moving electrical contacts. The symbols for the solenoid are listed in figure 16-6. Notice that they are very similar to the relay coil symbol but there are not any contacts.

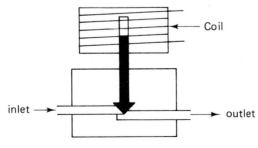

Figure 16–5   Solenoid valve

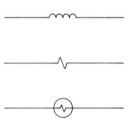

Figure 16-6   Solenoid symbols

Solenoid valves are used in air conditioning and refrigeration systems to control electrically such refrigerant valves as hot gas defrost valves in commercial refrigeration systems and reversing valves in heat pumps.

The coil in the solenoid must be supplied with the exact voltage listed on its data plate. It will react like a relay coil and not move the valve if too little voltage is supplied and it will quickly burn out if too much voltage is used. Use the following procedure to test and operate the solenoid valve on your trainer. Check off the steps as you complete them.

1. Turn off the power to your trainer.

2. Locate the solenoid on your trainer and record its required voltage____ _____.

3. Notice that the solenoid has only two wires on its coil. Set your ohmmeter to R × 10 and measure the resistance of your solenoid's coil. (The reading should be between 1K and 20K ohms.)

    a. If the reading is infinity (∞) change your meter to R × 100K and retest.

    b. If the reading is still infinity (∞) the solenoid coil is bad. Ask your instructor for a new one and repeat step 3.

4. Connect your solenoid wires to the proper voltage source.

    a. If it is 24 volts use R and C on the low voltage transformer.

    b. If it is 120 volts use L_1 and neutral in the single phase disconnect.

    c. If it is 230/208 volts use L_1 and L_2 in the single phase disconnect. **Notice!!**   Always apply power to a solenoid coil when it is on the valve or it will burn out because the valve helps hold down the current through the coil.

5. Apply power and listen for a click of the valve opening or closing.

6. If your solenoid valve has a good resistance reading, but it does not click, the valve is probably stuck. Turn the power off and on several more times to free it up. If it still will not click, call your instructor for help.

7. Turn off power and return all parts and wiring to their proper places.

This completes Unit 16. When you are ready, ask your instructor for the quiz on Unit 16.

17 Temperature Operated Controls

Tools Required

Voltmeter (0-250 volts scale)

Screwdriver (flat blade ¼ in.)

Clamp-on ammeter (30 amp capacity)

Needle-nose pliers (3 in. with insulated handles)

Lamp with 100 watt bulb

Heat gun, 100 watt

Objectives for Unit 17

At the conclusion of this unit the student should be able:

To explain the operation of the vapor pressure type control orally or in writing to the instructor's satisfaction.

To explain the operation of the bimetal type control orally or in writing to the instructor's satisfaction.

To test correctly the operation of the bimetal and vapor pressure type operators.

To explain the operation of the Klixon warp type switch orally or in writing to the instructor's satisfaction.

To test correctly the operation of the room heating and cooling thermostat.

To measure correctly the current flowing through the heat anticipator of the heating thermostat.

To explain the operation of the fan switch in the auto and on positions orally or in writing to the instructor's satisfaction.

To construct correctly a times-10 multiplier used to measure the heat anticipator current.

To explain the operation of the fan and limit switch orally or in writing to the instructor's satisfaction.

To test correctly the operation of the fan and limit switch.

TYPES OF TEMPERATURE CONTROLS

The most common temperature control for air conditioning and refrigeration equipment is the thermostat. The thermostat can turn on cooling equipment when the room temperature gets too warm and turn on heating equipment when the room temperature gets too cold.

Other types of temperature controls found in air conditioning and refrigeration equipment are: (1) the fan switch that turns on the furnace fan when the furnace warms up, and (2) the limit switch, which is a safety switch operated by temperature that will shut off the furnace if the temperature gets too high. On some equipment the temperature operated switch is located on the electrical panel and a remote bulb is extended into the conditioned space. This temperature control is called a remote bulb thermostat. Unit 17 will explain the operation, installation, adjustment, and troubleshooting of safety controls, and controlled switches used to protect equipment and operational controls used to operate equipment.

Most temperature operating and safety controls are made in two parts. The *switch* is the electrical part of the control and the *operator* is the part of the control that causes the switch to open or close.

TEMPERATURE OPERATORS

The two most widely used temperature operators are the *vapor pressure* type and the *bimetal* type. The vapor pressure type (see figure 17-1) has a bulb half filled with a liquid that boils easily, attached to a diaphragm chamber. As the temperature increases, the liquid in the bulb evaporates and the vapor causes the diaphragm to move inside the diaphragm chamber. The diaphragm's movement causes the switch to open or close.

The bimetal operator is made of the two metals with different expansion rates that have been pressed together (see figure 17-2(a)). As the temperature of the bimetal increases, one of the sides of the bimetal expands faster than the other causing the bimetal to bend or warp (see figure 17-2(b)). By bending the bimetal into a spiral the movement caused by the temperature change will be greater (see figure 17-2(c)).

TESTING THE SWITCH AND TEMPERATURE OPERATOR

The temperature control is made of two parts, the switch and the operator, that must work as a single unit. It is important that you know how to test both parts of the control. The switch part of the control may be as simple as a

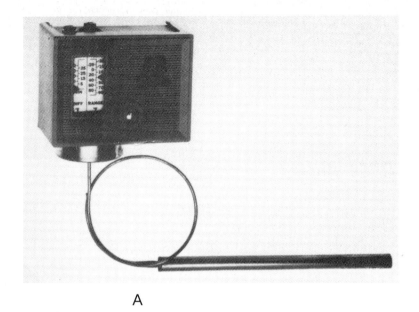

A

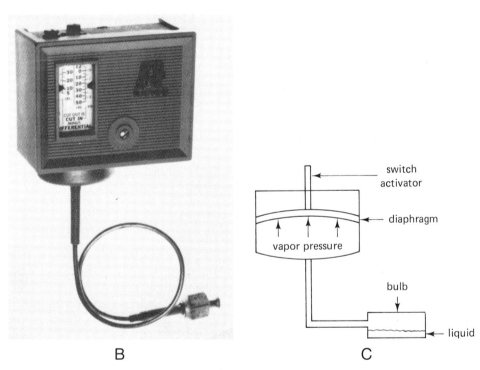

B

C

Figure 17-1 Vapor pressure controls (a) and (b) (courtesy of *Ranco Controls*)

single pole, single throw switch or as complex as a double pole, single throw switch.

Use the following procedure to test the switch part of the temperature control. You will need to test this part of the control *while* you are testing the operator. Check off the steps as you complete them.

1. Set the ohmmeter to R × 1 and zero.

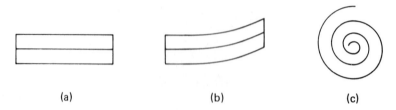

(a)   (b)   (c)

Figure 17-2   (a) Bimetal cooled; (b) bimetal heated; (c) bimetal spiral

2. Place one probe on each of the two switch terminals (use terminals R and Y for the room thermostat).

3. When the temperature control operator moves the switch, the ohmmeter reading should change from infinity (∞) to zero ohms if the switch is normally open (from zero to infinity if the switch is normally closed).

Use this procedure to test the bimetal operator for movement when the temperature changes. Check off the steps as you complete them.

1. Locate the room thermostat mounted on your trainer and remove its cover.

2. Notice it uses a spiraled bimetal operator. Do not touch the bimetal spiral since it is calibrated.

3. Your instructor will provide you with a lamp. Move the lamp near the bimetal. You should notice that as the spiral warms up it will try to unwind. (Use the three step procedure in the preceding paragraph to test the electrical part of the control at this time.)

4. Remove the lamp and you should see the spiral move in the opposite direction and wind up as it cools down. If the bimetal does not move on your thermostat, call your instructor.

5. Replace the cover on your thermostat and continue to the next test.

Some bimetal operators are disc shaped. These operators are sometimes called *warp switches* because the metal tends to bend or warp until it snaps as it is heated. This type of operator is usually calibrated to "warp" on and off at designated temperatures and is usually called a *Klixon*. (Klixon is the trade name of the most widely used warp swtich.)

Use this procedure to check the warp-type switch.

1. Locate the Klixon switch on your trainer.

2. Use the lamp to provide heat to the Klixon.

3. As the temperature reaches the temperature setting, you will hear the operator "snap" as it warps.

4. Remove the heat from the control and you will hear the switch snap again.

   a. If your switch will not snap, call your instructor.

The pressure vapor operator uses the liquid in the bulb to move a diaphragm. The liquid will have a predetermined boiling point. As more heat is added to the bulb, more pressure will build in the bulb and push against the diaphragm. The diaphragm will have atmospheric pressure (14.7 PSI) and spring pressure (usually adjustable) pushing back on the other side of the diaphragm. When the bulb pressure is greater than the atmospheric pressure and spring pressure, the diaphragm will move far enough to activate the switch. One advantage of this type of operator is that the switch can be mounted some distance from the bulb, usually on the control panel. This saves money and keeps the switch in a clean and dry location.

Use the following procedure to test the vapor pressure type operator on your trainer.

1. Locate the vapor pressure control on your trainer (see figure 17-1).

2. Remove the cover on your control.

3. Warm the bulb with your hand or the lamp. Notice that the switch will "click" when the temperature exceeds the setting on the switch. (Use the first test procedure in this unit—the three-step procedure—to test the electrical part of the control at this time.)

4. If the room temperature is warmer than the setting on the switch, use an ice cube to cool the bulb until the switch "clicks."

5. If the switch would not click when you added heat or cooled the bulb, call your instructor. The bulb for your control may have lost its charge. Ask your instructor to inspect the control.

6. Replace the cover on your control.

RESIDENTIAL HEATING THERMOSTAT

The room thermostat is the most widely used temperature control. The heating thermostat is an operational control, since it turns the heating off and on at specific temperatures. The temperature the thermostat operates at is called the *set point*. Heating equipment manufacturers have standardized the operation of the room thermostat. This means that if you understand the operation of one thermostat, you can figure out how other thermostats operate.

The electrical circuit for heating will be found between terminals R and W (see figure 17-3). The circuit consists of a mercury switch mounted

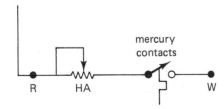

Figure 17-3   Heating thermostat

on the spiral bimetal operator (see figure 17-4) and an adjustable variable resistor called a *heat anticipator* (see figure 17-5).

The heat anticipator adds a small amount of heat to the bimetal operator during the heating cycle. This causes the thermostat to open the heating switch approximately 1° before the room reaches the set point. This allows the fan to move the large amount of heat built up in the furnace into the room after the fire is shut off without overheating the room. If the thermostat waited until the room warmed to exactly the set point before it open the circuit to the furnace, the excess heat left in the furnace would bring the room 2° to 3° above the set point.

MEASURING THE HEAT ANTICIPATOR AMPERAGE

The *heat anticipator* is a variable wire wound resistor that is in series with terminals R and W inside the thermostat. This means that any current going through terminal W must go through the heat anticipator causing it to heat the bimetal slightly. If too much current is pulled through terminal W, the thin wire of the heat anticipator will burn. This will cause an open in the circuit R to W and the furnace will not come on. Use the following procedure to measure the current going through the heat anticipator. Check off the steps as you complete them.

1. Locate the room thermostat on your trainer and remove its cover.

2. Locate the adjustable heat anticipator and set it to 1.4.

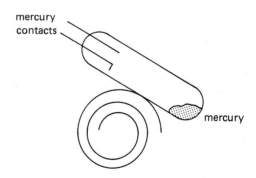

Figure 17-4 Spiraled bimetal

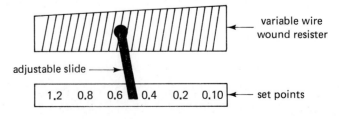

Figure 17-5 Heat anticipator

3. Turn the power off to the trainer and connect the following circuit. (The relay coil will represent the coil in the gas valve.)

4. Ask your instructor to provide you with 70 inches of single strand thermostat wire (#18 gauge).

5. Coil the wire around the palm of your hand so that you have exactly 10 coils. Leave approximately six inches on each end of the coil for terminals (see figure 17-6). Tape the coils in place with electrical tape. This coil will multiply the current going through it 10 times when used with your clamp-on ammeter.

Figure 17-6   Multiplier coil

6. Remove the wire from terminal W on the thermostat terminal board. Connect one end of your multiplier coil to the relay coil and the other end of your multiplier coil to terminal W (see figure 17-7). This will connect your multiplier coil in series with the heat anticipator.

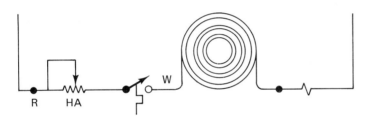

Figure 17-7   Multiplier coil in circuit

7. Connect your clamp-on ammeter around the multiplier coil and set the scale to the lowest range.

8. Energize the circuit and adjust the heating thermostat to its highest setting. The relay should close and the current draw should show on the ammeter.

9. Record the amperage reading_____.

10. Since you are using a times-10 multiplier coil, you must divide the ammeter reading by 10 to determine the actual current flowing through the heat anticipator.

    Ammeter Reading ÷ 10 = Actual Current

11. Record the actual current of your circuit_____.

12. Re-adjust the thermostat to the lowest setting and the relay will be de-energized. Notice that the current flowing through the heat anticipator is now zero. This shows that the heat anticipator has current

flowing through it only when current is going to the gas valve and the furnace is on.

13. Adjust the heat anticipator to the same number as the amount of current you measured in step 10.

14. Energize the circuit again. Notice that the current has remained the same. Only the amount of heat going to the bimetal from the heat anticipator has changed. Turn off the power and leave the circuit in place.

   Note: Remember the heat anticipator is used to match the heating equipment to the house for comfort. It will keep the furnace from getting the house too warm during the heating cycle by shutting off the heat approximately 1° before the set point. The setting on the heat anticipator may be higher than the actual current flow, but never lower, as it may burn out the anticipator. The final setting of the heat anticipator will be to the customer's comfort.

COOLING THERMOSTAT WITH FAN SWITCH

The room cooling thermostat is used to turn the air conditioning system off and on at the set point temperature selected by the customer.

The cooling thermostat has a mercury bulb switch that is mounted on the spiral bimetal operator. This switch completes the circuit between R and Y. R is the "hot" or supply terminal for the thermostat; terminal Y is the cooling terminal.

The other part of the cooling thermostat is the fan switch. The fan switch is a single pole, double throw switch (see figure 17-8). The single pole is connected to terminal G. Terminal G is called the fan terminal on the thermostat. When the switch is placed in the ON position, terminal G gets power from terminal R. When the fan switch is placed in the AUTO position, terminal G gets power from terminal Y.

The ON position is used when the customer wants to run the furnace fan without turning on the air conditioning system. This will simply circulate

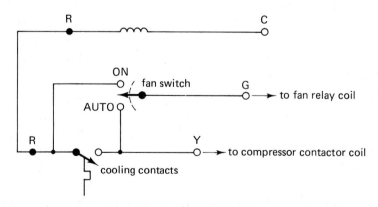

Figure 17-8   Fan switch on cooling thermostat

air in the home. When the fan switch is in the AUTO position the furnace fan will only come on when the air conditioning system is brought on by the bimetal changing and the mercury switch closing.

Use the following procedure to test your cooling thermostat and fan switch. Check off the steps as you complete them.

1. Turn off the power to the trainer.

2. Use the diagram below and wire the following circuit. You will need two relays with 24 volt coils.

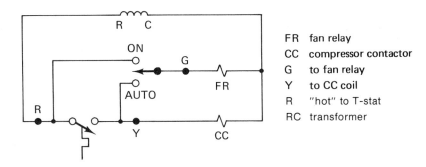

FR fan relay
CC compressor contactor
G   to fan relay
Y   to CC coil
R   "hot" to T-stat
RC transformer

3. Set thermostat to air conditioning and adjust the temperature setting to the highest number. This will keep the air conditioning system off. The relay marked CC will not be energized.

4. Turn the power on and turn the fan switch to the ON position. Notice that the fan relay closes every time you turn the fan switch to ON.

5. Turn the fan switch to AUTO. Notice that the fan relay now opens. The reason the fan relay turned off when you turned the fan switch to the AUTO position is that the air conditioning system is turned off, and when the fan switch is in the AUTO position terminal G gets power from terminal Y.

6. Leave the fan switch in the AUTO position. Carefully remove the cover on the thermostat so you can see the mercury bulb.

7. Adjust the cooling set point to the lowest number. Notice the movement of the mercury bulb. When the mercury makes contact at the opposite end, it will complete the thermostat circuit from R to Y (see figure 17-9). The compressor contactor (CC) and the fan relay should both be energized now.

8. Move the fan switch from auto to ON. Notice that the fan relay stays closed. The fan relay stays energized because terminal G is supplied power through the fan switch in both position when terminal Y is "hot." When terminal Y is energized, the thermostat is said to be "calling for cooling." Turn the power off and leave this circuit in place.

Some room thermostats, like the one on your trainer, have the heating and cooling controls mounted on one base. This type of thermostat is called a *heat/cool thermostat.* The heat/cool thermostat is tested as explained in the previous tests by checking each circuit independently.

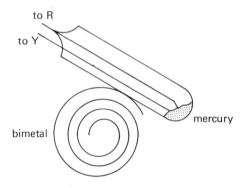

Figure 17-9   Cooling contacts

THE FAN AND LIMIT CONTROL

The warm air furnace does *not* use a fan relay and terminal G on the thermostat to control the furnace fan. Instead, a fan and limit switch is used (see figure 17-10). This control has two separate switches mounted on one temperature operator. This control uses a long spiraled bimetal operator that is placed where it will sample the air temperature leaving the furnace heat exchanger.

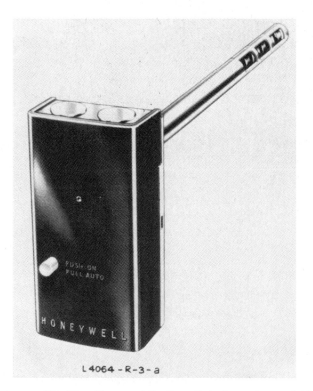

Figure 17-10   Fan and limit switch (courtesy of *Honeywell Inc.*)

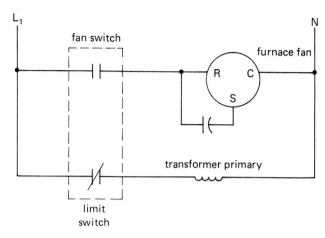

Figure 17-11 Fan and limit switch circuit

Half of the control, the fan switch, turns the furnace fan on when the air temperature reaches approximately 120°F. This will ensure that only warm air will come out of the registers. After the fire is turned off the heat exchanger in the furnace will cool down. When the temperature drops to approximately 90°F, the fan switch turns the furnace fan off. The set points on the fan switch are usually adjustable.

In figure 17-11 you can see that the fan switch controls 120 volts for the furnace fan and is rated for 15 amps.

The second part of the fan and limit control is the limit switch. The limit switch operates from the same bimetal as the fan switch and will control the maximum temperature of the heat exchanger. From figure 17-10 you can see the limit switch is a normally closed switch that will control either the transformer primary voltage or the transformer secondary voltage going to the gas valve. In either case the gas valve will be de-energized if the heat exchanger's temperature gets too warm. The main reason the heat exchanger will overheat on a furnace is from insufficient air moving across it. This may be due to an inoperative furnace fan, dirty furnace filter, or closed registers.

Use the following procedure to test the fan and limit control on your trainer. Check off the steps as you complete them.

1. Turn the power off on your trainer.

2. Locate the fan and limit switch on your trainer.

3. Set your ohmmeter to R × 1 and zero.

4. Remove the cover from the fan and limit switch and locate the fan part of the switch.

5. Test the fan switch with the ohmmeter. Notice that it is a normally open switch.

6. Your instructor will provide a heat source to warm the bimetal. (Use a heating gun or lamp capable of producing 120°.)

7. Notice the switch closes at the set point. (If not, call your instructor.)

8. Place the ohmmeter across the limit switch. Notice that this is a normally closed switch.

9. Apply heat to the switch bimetal again. Notice that at the set point temperature the limit switch opens.

10. Connect the following circuit.

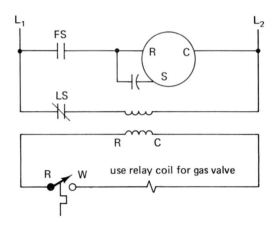

11. Apply power to the trainer and notice that the fan motor is not running. Explain why. _____
    _____
    _____
    _____

12. Set the thermostat to the highest number. The (relay) gas valve coil should energize.

13. Is the fan running? Why?_____
    _____
    _____
    _____

14. Apply heat to the bimetal. What happens when the temperature exceeds the fan switch set point? _____
    _____
    _____
    _____
    _____

15. Turn the thermostat to the lowest number (heat is turned off).

16. What happens to the fan when the bimetal cools down? _____
    _____
    _____
    _____
    _____

17. Turn the thermostat to the highest number (heat is on). Apply heat to the bimetal until the limit switch set point is exceeded.

a. What happens to the gas valve (relay coil) when the limit switch is exceeded?_____

_____

_____

_____

b. Does the fan turn off? Why? _____

_____

_____

_____

18. Turn off the power to the trainer and remove all wires and place them in their proper location.

CONTACT RATING FOR TEMPERATURE CONTROLS

The temperature controls you will find in air conditioning and refrigeration equipment are used as operational and safety controls. All temperature controls have this in common: that the change in temperature makes the operator move, engaging the electrical switch which energizes the system. The electrical switch must be examined to determine the contacts size and rating. The voltage and current rating on the switch *can not be exceeded*. For example, if you were to try to control a 120 volt circuit through a room thermostat rated for 24 volts, the excessive voltage would cause the mercury in the switch to boil and explode the glass tube.

Use the following procedure to check the contact rating for temperature controls on your trainer.

1. Turn off power to your trainer.

2. Locate the following temperature controls on your trainer: room thermostat, remote bulb thermostat, and Klixon warp switch.

3. The room thermostat will normally be rated for 24 volt circuits unless otherwise specified. Remove the control's cover where needed and locate the switch ratings.

4. Record the voltage and current rating of each switch below.

	Volts	Amps
Remote bulb thermostat	_____	_____
Klixon warp switch	_____	_____
Fan and limit switch	_____	_____

5. Replace the covers you removed in step 3.

6. Replace all wires to their proper locations.

This completes Unit 17. Ask your instructor for the quiz on Unit 17 when you are ready.

18 Pressure Controls

Tools Required

Ohmmeter (R × 1 scale)
Screwdriver (flat blade ¼ in.)
Manifold gauge set (high and low pressure gauges)
Refrigerant tank with R-22

Objectives for Unit 18

At the conclusion of this lesson the student should be able:

To explain orally or in writing the operation of a pressure control to the instructor's satisfaction.

To test correctly the operation of a preset low pressure control or high pressure control to check the set point.

To adjust cut-in and cut-out set points correctly for a low pressure switch.

To explain orally or in writing the operation of the oil pressure control to the instructor's satisfaction.

To connect an oil pressure switch correctly and check its operation.

To identify correctly the electrical symbols for low pressure, high pressure and oil pressure control.

PRESSURE CONTROLS

Air conditioning and refrigeration systems must be protected against very high pressure and very low pressure. Some systems use pressure controls to turn the unit off and on at the right temperature since refrigerant pressure at any temperature can be predicted. This unit will explain the operational theory, installation, adjustment, and troubleshooting of pressure controls.

OPERATION OF THE PRESSURE CONTROL

The pressure control (see figure 18-1) is similar to the temperature control in that it has two main parts: the diaphragm operator and an electrical switch. The diaphragm is fitted into a chamber and moves as the pressure changes. On one side the diaphragm chamber is connected directly into the refrigeration cycle with a flare fitting. In this way the diaphragm in the control can sense the refrigeration system's pressure directly. On the other side, the diaphragm is conneced to the cut-in/cut-out electrical switch. It is thus activated by the refrigeration system's pressure.

The set point (pressure at which the diaphragm opens or closes the switch) is determined by spring tension against the diaphragm. The spring tension is adjustable on some controls and preset on others. Use the following procedure to test the operation of the pressure control's switch and and diaphragm. Check off the steps as you complete them.

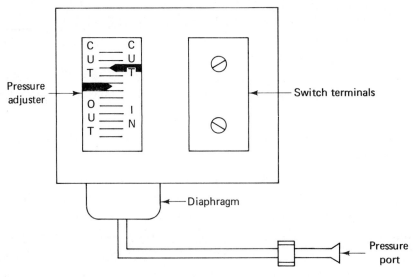

Figure 18-1   Pressure control

1. Turn off the power to your trainer.

2. Locate the pressure control on your trainer. Use figure 18-1 as reference. (Pressure controls can be easily identified by the flare fitting connected to their diaphragms.)

3. Remove the cover from your pressure control.

4. Set the ohmmeter on R $\times$ 1 and zero.

5. Locate the switch terminals on the pressure control. If there are two terminal screws it is a single pole, single throw switch. If the control has four screws it is a double pole, single throw switch (see figure 18-2).

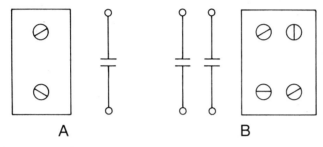

A   B

Figure 18-2 (a) Single pole, single throw switch; (b) double pole, single throw switch

The single pole switch is used on 24 volt and 120 volt systems. The double pole switch is used on 230 volt systems.

6. Test the switch terminals to determine if they are normally open or normally closed.

   a. If the meter measures zero resistance, the switch is normally closed.

   b. If the meter measures infinite (∞) resistance, the switch is normally open. (In this case normally open and normally closed refer to the switch with no pressure applied to the diaphragm.)

   c. If the switch on your pressure control is normally open, you have a low pressure control.

   d. If the switch on your pressure control is normally closed, you have a high pressure control. When pressure increases past the set point, the contacts will open.

7. Your instructor will provide you with a tank of Refrigerant 22 and a set of manifold gauges.

   a. Remove the low pressure (blue) hose from the manifold set.

   b. Connect the flare fitting of your pressure control to low side of the manifold set where you removed the hose.

Caution: You must be very careful when handling the refrigerant tank. *Liquid* refrigerant will freeze any part of your body it touches. This means, for example, that if you get it in your eyes it will cause permanet vision loss. For this exercise, keep the tank upright so that the valve is on top, and so that only vapor will be allowed to be released from the tank.

   c. Connect the middle hose of your manifold gauge set to the refrigerant tank.

8. Open the valve on the refrigerant tank.

9. Open the low side valve. (The refrigerant vapor should flow to the switch diaphragm.)

   a. When the gauge shows the pressure has exceeded the control's set point, the switch should change from normally open to closed or from normally closed to open. (You may have to lower the set point on the high pressure switch to be able to activate it with tank pressure.)

   b. Close the tank's valve.

   c. Release the pressure from the control slowly. Notice the switch will "click" or change positions as the pressure drops below the set point.

   d. Place the ohmmeter across the switch contacts and pressurize the control again. This time, observe the ohmmeter as you hear the switch click. What does the ohmmeter show?_____

10. Close the tank's valve and release the pressure from the control. What does the ohmmeter show as the pressure drops below the set point? _____

11. Leave the refrigerant tank connected to the pressure control and continue to the adjustment of the pressure control.

ADJUSTMENT OF THE PRESSURE CONTROL

Some pressure switches are adjustable. Adjustable switches usually have a dial and adjustment screws. Some pressure controls are preset at the time of manufacture and cannot be adjusted (see figure 18-3).

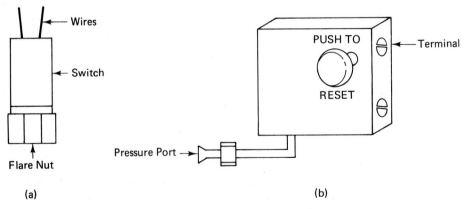

(a) (b)

Figure 18-3 Nonadjustable switches

Controls that cannot be adjusted only need to be tested as in the previous procedure to be sure the control operates at the design set point.

The adjustable high-pressure switch will have one adjustment for the maximum allowable pressure. If the operating air conditioning or refrigeration system's pressure exceeds the maximum allowable pressure, the high pressure control will open its switch and shut off the compressor.

The low pressure switch will have two adjustments: cut-out pressure and cut-in pressure. These are sometimes referred to as cut-out and differential switches. The cut-out is the pressure that the low pressure switch will open.

The low pressure switch can be used to control the temperature in a commercial refrigeration system like a walk-in cooler or frozen food case. (This is possible because the pressure of a refrigerant at any given temperature can easily be determined from a pressure-temperature chart.) The low pressure control must have a cut-in pressure and cut-out pressure. If the pressure switch operated at one set point, the unit would *short cycle* or shut off and turn on again too soon. The pressure differential between cut-in and cut-out allows the refrigerated space to increase in temperature a few degrees before the system comes on again. In other words, the temperature of a refrigeration system would be a range of temperatures; for example 18° to 21°, or 36° to 39°. Use the following procedure to set the differential or cut-in and cut-out set point on the low pressure control. Check off the steps as you complete them.

1. Use the previous setup with gauge manifold set connected to the refrigerant tank and low pressure control.

2. Set the range or cut-in pressure set point to 30 PSI.

3. Adjust the cut-out to 25 PSI. This makes a differential of 5 PSI.

4. Open the tank valve and gauge manifold valve to allow 60 PSI into the low pressure control.

   a. Check the switch contacts with your ohmmeter. Notice that they are closed.

   b. Turn off the refrigerant tank valve and allow the pressure to slowly bleed off until the switch opens. Record the pressure that the switch opens_____. This is the cut-out pressure.

   c. Open the refrigerant tank valve and slowly allow pressure to build until the switch closes. Record the cut-in pressure_____

   _____.

   d. Subtract the cut-out pressure from the cut-in pressure. This is the differential_____. (If your cut-in and cut-out or differential are different from what you set, call your instructor.)

5. Change the cut-out to 20 PSI. This changes the differential to 10 PSI. Repeat step 4 to test your low pressure control.

6. Your instructor will provide you with a pressure-temperature chart.

   a. Find the temperature for R-22 that your low pressure control would cut-out at 20 PSI_____

b. Find the temperature for R-22 that your low pressure control would cut-in at 30 PSI _____

c. What would the cut-in and cut-out temperatures for (a) and (b) of question 6 if the refrigerant used was R-12?

Cut-in_____ Cut-out_____

7. Close off the valve on the refrigerant control and remove the low pressure control from the manifold set.

OIL PRESSURE CONTROL

The oil pressure control (see figure 18-4) is used on air conditioning and refrigeration systems that have compressors using a pressure-type oil pump. The oil pressure will shut off the compressor if the oil pump does not provide pressure.

The oil pressure control works from a pressure differential. This means that pressures from two separate sources are checked against each other. In figure 18-4 notice that two ports are used to sample low side (or crankcase) pressure and oil pressure. The bottom port samples oil pump pressure. The top port samples low pressure. When the oil pump pressure is a minimum of 15 PSI larger than the low side pressure, the pressure sensing element will move upward and the magnetic follower will move with it. The magnetic follower will activate the switch.

The other part of this control is a time delay clock. The time delay clock is a solid-state time delay circuit. Some older controls use a heating element and bimetal switch or mechanical clock. The oil pressure control needs a time delay to give the oil pump in the compressor 120 seconds to begin pumping. When the compressor is not running, the control switch will be off because the low pressure port will have approximately 110 PSI for R-22 and 76 PSI for R-12 due to the ambient temperature of 70°. Since the oil pump is not turning, the oil pressure is zero. When the compressor begins to turn the oil pump, the oil pressure will begin to build and the system low-side pressure will drop. When the oil pressure exceeds the low pressure, the switch will close. If the oil pump does not produce the 15 pounds pressure differential within 120 seconds (2 minutes), the time delay circuit opens the compressor circuit and shuts off the compressor. Use the following procedure to test the oil pressure control. Check off the steps as you complete them.

1. Turn off power to your trainer.

2. Locate the oil pressure control and remove its cover.

3. Use the diagram in figure 18-5 to connect your compressor to the oil pressure switch. Since your trainer is a hermetic compressor it does not have an oil pump. For this exercise, do not connect anything to the two pressure ports.

4. If your compressor is 230 volts, use a relay with a 230 volt coil and connect the circuit as it is drawn.

A

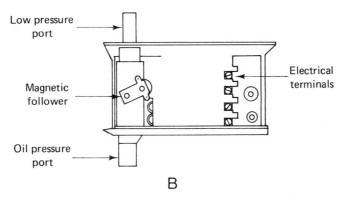

Low pressure
port

Magnetic
follower

Electrical
terminals

Oil pressure
port

B

Figure 18-4   Oil pressure control (courtesy of *Ranco Controls*)

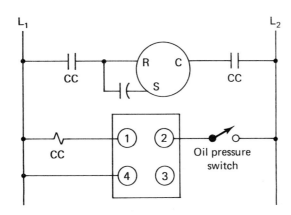

Figure 18-5   Oil pressure control in circuit

5. If your compressor is 120 volts, use a relay with a 120 volt coil and use neutral in place of L_2 and terminal 3 instead of terminal 2.

Notes:

a. Between 1 and 2 is the set of oil pressure control contacts.

b. Between 4 and 2 is the time delay element.

c. Use terminal 3 in place of terminal 2 for 120 volt control systems.

d. The on/off control is a 230 volt rated switch. It could be a low pressure or temperature controller used to control the conditioned space temperature.

6. When you have completed wiring the circuit, call your instructor to check it before you apply power.

7. Apply power to your circuit. Since there are no pressure lines connected to the control it will run for approximately 120 seconds, and then the compressor will be shut off by the oil pressure contacts opening the relay's coil circuit.

a. Use a watch to record the amount of time the system runs. This is the time delay_____.

8. To start the system again the oil pressure control must be manually reset by pushing in the button on the front. This time hold the magnetic follower in the up position. This will simulate a proper oil pressure condition. Hold the magnetic follower up for three to four minutes and then release it.

a. Notice that while the magnetic follower is up (simulating proper oil pressure) the compressor continues to run.

b. Notice that when the magnetic follower is dropped (simulating a loss of oil pressure) the compressor stops after the short time delay.

9. Reset the oil pressure control and hold the magnetic follower up. Allow the system to run for four minutes and turn off the on/off control switch.

a. Notice that the compressor shuts off immediately.

b. Turn the on/off control on again. Notice that the reset button did not have to be reset this time. The reset needs to be pushed in only when low oil pressure shuts off the unit. This causes a technician to check out the system when the reset is pushed. The system can cycle on and off normally on the temperature control as long as the oil pressure is adequate.

10. Turn off power to the circuit and remove all wiring.

ELECTRICAL SYMBOLS FOR PRESSURE CONTROLS

The following symbols are used for the pressure controls. The *low* pressure control opens when pressure drops, and it closes when pressure rises. This is indicated by having the switch symbol below the terminal

Low Pressure Control

The ⌐⌐ symbolizes a switch which is operated by pressure.

The *high* pressure switch opens when pressure rises and closes when pressure fails. The switch symbol is above the terminal.

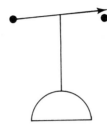

High Pressure Control

Sometimes the low pressure switch is indicated by a set of normally closed contacts marked L.P., the high pressure switch is indicated H.P., and the oil pressure switch is indicated O.P. (see figure 18-6).

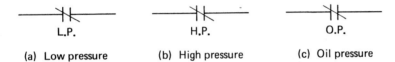

L.P.	H.P.	O.P.
(a) Low pressure	(b) High pressure	(c) Oil pressure

Figure 18-6 Electrical symbols

This completes Unit 18. Ask your instructor for the quiz on Unit 18 when you are ready.

19 Timer Controls

Tools Required

Voltmeter (250 volt scale)
Ohmmeter (R × 1K scale)
Screwdriver (¼ in. flat blade)

Objectives for Unit 19

At the conclusion of this unit each student will be able:

To explain orally or in writing the operation of a time clock timer control to the instructor's satisfaction.

To wire a time clock to a circuit correctly and set its timer to given specifications.

To test correctly a timer clock for correct operation.

To explain orally or in writing the operation of a heating sequencer to the instructor's satisfaction.

To wire a set of heating sequencers correctly and check operation.

To test a heating sequencer correctly with an ohmmeter.

TIMER CONTROLS

Air conditioning and refrigeration systems need timers to keep track of defrost cycles. Electric furnaces need time delays of several minutes to stage heating elements. This unit will explain the operation, installation, adjustment, and troubleshooting of time controls and sequencers.

OPERATION OF TIMERS

The timer shown in figure 19-1 has a 24 hour clock and is used to control defrost cycles in commercial refrigeration systems. It is also used to turn off large air conditioning systems when buildings are unoccupied at night or on weekends, and turn the system back on before the start of business the next day.

From the diagram in figure 19-2 you can see that the timer has two separate parts; the clock motor and two sets of contacts.

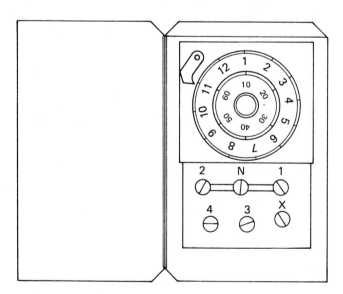

Figure 19-1   Time clock

The timer motor is connected to 120 volts between terminals X and N and has a very constant speed. The open contacts are between 3 and 1 and are connected to line 1. The closed contacts between terminals 2 and 4 are also connected to the line 1.

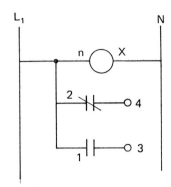

Figure 19-2 Typical timer electrical diagram

The clock can be set to change the open and closed contacts for any duration of time at any specific time of day or night. For instance, the motor might open the closed contacts (turning off the air conditioning) at 9:30 P.M. after everyone has left a store and remain open for 10 hours, closing again at 7:30 A.M. to pre-cool the store before it opens at 10:00 A.M. Another example would be to close the open contacts every 12 hours for a 10 minute defrost cycle on a commercial refrigeration system.

INSTALLING AND TESTING THE TIMER CONTROL

As an air conditioning and refrigeration technician, you must be able to install, adjust, and test the timer control to be sure it is operating correctly. Use the following procedure to install, adjust and test the timer control. Check off the steps as you complete them.

1. Turn off the power to the trainer.

2. Locate the timer control on your trainer.

3. Open the door and connect the following circuit to the terminals indicated.

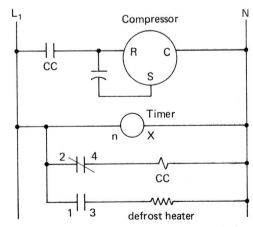

Figure 19-3 Time clock in circuit

a. Connect L_1 to n

b. Connect X of the timer to N of disconnect (use L_2 if 230).

c. Connect L_1 to CC contacts.

d. Connect contacts of CC to R of compressor.

e. Connect the run capacitor to R and S of compressor.

f. Connect the compressor terminal C to N (use L_2 if 230).

g. Connect N to CC coil.

h. Connect the timer terminal 4 to CC coil.

i. Connect N to defrost heater.

j. Connect the other side of the heater to terminal 3 of timer.

Call your instructor to approve your wiring.

Instructor's approval _____

4. Set the pins for the time of day the defrost cycle is to occur (see figure 19-4). In this case we want defrosting to occur at 6:00 A.M. and 6:00 P.M.

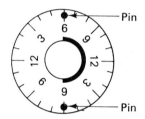

Figure 19-4   Time clock set points

5. Set the length of the defrost cycle to three minutes. (Refrigeration equipment manufacturers will usually specify the length of the defrost cycle according to application.)

6. Adjust the timer to just before 6:00 A.M. or 6:00 P.M.

7. Call your instructor to check your adjustment.

Instructor's approval _____

8. Apply power and observe the timer until the defrost cycle occurs. The compressor should be on and the heater should be off at this time.

9. When the defrost cycle starts, the compressor should shut off and defrost heater should turn on for three minutes. At the end of the three minute defrost cycle, the compressor should come back on and the heater should go off.

a. When will the next cycle occur? _____

b. What would you do to change the defrost cycle to 10 minutes?____
_____

c. What would you do to change the interval between cycles to six hours? _____
_____

10. To synchronize the timer to the real time, rotate the large dial to the exact time it is now. For example, if it is 10:00 A.M. set the dial as shown in figure 19-5.

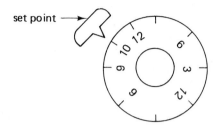

set point

Figure 19-5   Setting timer dial

11. To test the timer to be sure it is operating, test terminals X and n for 120 volts.

    a. If you do not have 120 volts at terminals X and n, the circuit providing power to the clock is defective or you have lost supply power.

    b. If the voltage between X and n is 120 volts and the timer gears are not moving, you can check between X and n for resistance (approximately 1K ohms). Replace the timer if you have 120 volts at X and n *and* the time gears do not move.

12. Turn off all power at the trainer and return all parts and wires to their proper places.

ELECTRIC HEAT SEQUENCER

The electric heat sequencer is also a time control (see figure 19-6). It generally provides delays of two to four minutes. It also allows electric heating coils to be staged so that the first stage of electric heat comes on before the second or third stage. Usually, the first stage of electrical resistance heat is the smallest of the three stages. For example, stage one may use 5KW (or 17,060 BTU) and stages two and three may use 10 KW (or 34,120 BTU) each.

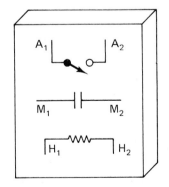

Figure 19-6   Heating sequencer

In locations where customers have a *demand* type electric meter it is useful to allow only the smallest bank of electric heat to come on until the house needs more than 17,000 BTU. Then the second stage will come on and the sequencer can keep the third stage off until it is needed.

OPERATION OF THE HEATING SEQUENCER

From the diagram of the heating sequencer in figure 19-7, you can see it has a heater and two sets of normally open contacts. The heater between H_1 and H_2 is rated for 24 volts and provides the heat to operate the bimetal switch that closes the contacts. The heat from the resistor takes two to four minutes to make the bimetal move to close the contacts. This provides the time delay for the sequencer. When current is removed from the heater it will take two to four minutes for the bimetal to cool and open the contacts again. In this way the sequencer provides time delay to the off cycle. This works well when the sequencer is controlled by a thermostat with a heat anticipator.

The two sets of contacts, the main and auxiliary, have different current ratings. The main contacts between M_1 and M_2 are rated for 25 amps and the auxiliary contacts between A_1 and A_2 are rated for 1 amp. For this reason, the main contacts are used to control the 5 KW and 10 KW heaters. The auxiliary contacts can only handle small control currents.

Use the following procedure to test the operation of the heating sequencer. Check off the steps as you complete them.

1. Turn off power to your trainer.

2. Locate the heating sequencer on your trainer.

3. Set your ohmmeter on R $\times$ 100K and zero.

4. Check the main contacts M_1 - M_2 and auxiliary contacts A_1 to A_2 to be sure they are open. Both resistance tests should read infinity. If your reading is other than infinity, the contacts are not completely open. This could be caused by a warped bimetal. If this is the case, call your instructor for a new sequencer.

5. Test the heater from H_1 to H_2, for continuity. It is usually less than 1 K. If the reading is infinity on the 100K scale, the heater is open and the sequencer must be replaced.

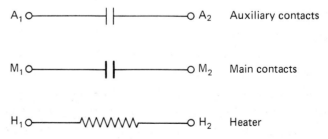

Figure 19-7   Electrical diagram of heating sequencer

6. Connect the sequencer as shown in figure 19-8.

   Note: Connect the transformer to the proper primary voltage listed on your transformer.

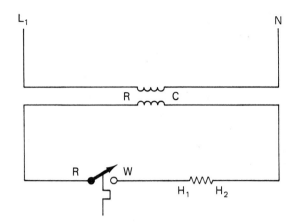

Figure 19-8   Diagram of heating sequencer in circuit

7. Be sure the heat anticipator on your thermostat is set to 1.4. If it is set lower than that, the current used by the heater may burn out the heat anticipator.

8. Apply power to H_1 and H_2 by turning on the primary voltage and setting the thermostat to the highest setting for heating.

9. Check the voltage at H_1 and H_2 record_____.

   a. If the voltage is higher than 25 volts, turn the power off and call your instructor.

   b. If the voltage reading is zero, check your circuit to determine where the loss of voltage has occurred.

10. Set your ohmmeter to $R \times 1$ and place the leads on M_1 and M_2. Record the time it takes the contacts to close once power has been applied to H_1 and H_2. The time delay is caused by the bimetal being warmed slowly by the heater.

11. After the contacts have closed, check the auxiliary contacts A_1 and A_2 to be sure they have also closed.

12. Turn the power off to H_1 and H_2 by setting the thermostat to the lowest number.

   a. Leave the ohmmeter across M_1 and M_2.

   b. Record the time it takes the contacts to open. (This is called the *off delay*.)_____

13. Turn off power to the trainer and wire the next circuit.

   Note: Use a 115 volt lamp in place of the heater strips. Your trainer does not have a large enough current capacity to operate 5 KW heaters.

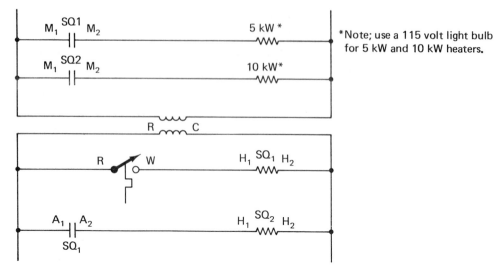

*Note; use a 115 volt light bulb for 5 kW and 10 kW heaters.

This circuit is similar to the circuit you will find in most electric furnaces. The circuit will operate in this way: When the power is turned on at the disconnect, only the transformer will be operating, supplying 24 volts at R and C. When the thermostat closes between R and W the heater of the first sequencer SQ1 will be powered. After two to four minutes its main and auxiliary contacts will close. The main contacts will provide power to the heater of the second sequencer SQ2. After two to four minutes the next set of heat, 10 KW, will energize. When the thermostat is satisfied (the room has warmed up to the set point) the R to W switch will open and power will be removed from the heater of SQ1. After two to four minutes, the first bank of heat (light bulb 1) will shut off.

The power to the heater of SQ2 will also be interrupted at this time because the auxiliary contacts of SQ1 will open, too. After two to four minutes the second set of heat (light bulb 2) will turn off as the bimetal of SQ2 cools and opens its main contacts.

Complete the wiring of this circuit and call your instructor to approve your wiring prior to applying power_____.

1. Apply power to the disconnect and set the thermostat to the highest setting.

2. Keep track of the time it takes for the first light bulb to turn on_____
    _____.

3. Keep track of the time it takes for the second light bulb to turn on once the first light comes on_____.

4. Set the thermostat to the lowest setting and record the time for the first light to turn off_____.

5. Record the time for the second light bulb to turn off_____
    _____.

6. Predict what would happen to your circuit if the heater of SQ1 went bad.
    _____
    _____
    _____

a. Remove the wire connecting H_2 of SQ1 to C and check your results against your prediction in number 6.

7. Predict what would happen if the heater of SQ2 went bad and the heater of SQ1 still functioned. _____
_____
_____
_____

    a. Remove the wire connecting H_2 of SQ1 to C and check your results against your prediction in number 7.

8. Predict what would happen if the main contacts of SQ1 went bad._____
_____
_____
_____

    a. Remove the wire from M_2 of SQ1 to light bulb 1 and check your results against your prediction in number 8.

9. Turn off power to your trainer. Remove all wiring and place it in its place.

    This completes Unit 19, Timers and Sequencers. Call your instructor for the quiz on Unit 19 when you are ready.

20 Overcurrent Controls

Safety for Unit 20

At times the circuits and disconnects you will be working with will be powered with up to 230 volts. You must be aware of and take appropriate safety precautions relative to electrical shock hazards and wear safety glasses while working.

Tools Required

Voltmeter (250 volt scale)
Ohmmeter (R $\times$ 1K scale)
Screwdriver (¼ in. flat blade)

Objectives for Unit 20

At the conclusion of this unit the student should be able:

To explain orally or in writing the operation of a dual element fuse.

To explain orally or in writing the operation of a circuit breaker.

To explain orally or in writing the operation of an internal overload.

To select the correct fuse, fuse disconnect, and wire size for an operating system.

To explain orally or in writing why a single element fuse cannot be used to start a motor.

OVERCURRENT CONTROLS

Air conditioning and refrigeration systems must be protected against overcurrent. The air conditioning and refrigeration technician must be able to select the proper size control to be sure the system is adequately protected against overcurrent. This unit will explain the operation and selection of overcurrent control devices such as fuses, circuit breakers, and overloads.

OVERCURRENT PROBLEMS

There are two basic types of overcurrent against which a system must be protected. One is the short circuit where L_1 or L_2 touch each other, neutral, or the cabinet ground. In this case the current will surge out of control since circuit resistance will be near zero. In many cases, short circuit current can reach 10,000 amperes or more if not cut off. Short circuit current can easily start a fire or destroy the disconnect, wire, and other electrical components if the circuit is not interrupted (opened) quickly.

The second type of overcurrent is callled a *slow overcurrent* or overload. This occurs when a motor or other electrical load draws more current than it is rated for. This could happen in a hermetic compressor if it tries to pump the refrigerant as a liquid, or in a fan motor if its bearings were not lubricated. The current in this case could exceed the normal amount by as little as 20% for several hours. Such an overcurrent results in severe overheating of the motor winding until its insulation is destroyed and the motor becomes inoperative.

FUSES

There are two basic types of fuses used today: the *single element fuse* and the *dual element fuse*. The single element fuse consists of a conducting link or fuse element that will melt and separate at a predetermined temperature. Heat builds up on the link when the fuse carries too much current. The excessive heat melts the link and the circuit is interrupted, stopping the current flow (see figure 20-1).

If a single element fuse is used to protect a motor, a basic problem must be overcome. That is the problem of starting amperage. For instance if the motor has a full load amperage rating (FLA) of 10 amps, it will be protected by a 10 to 15 amp fuse. When the motor starts, its locked rotor amperage (LRA)

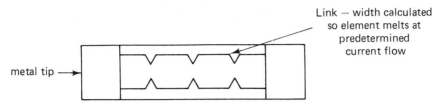

Figure 20-1　Single element fuse

will be three to five times the FLA. This means the LRA will be 30 to 50 amps for the motor. If the 15 amp single-element fuse is used, it will open and interrupt the circuit every time the motor starts. If a 30 or 50 amp fuse is used, the motor will be able to start, but the fuse will be too large to protect the motor in case of a bad bearing or other minor overload where the FLA may only increase to 20 amps. For this reason, the single element fuse is not capable of providing overcurrent protection for motors.

The dual element fuse has two parts or elements to provide protection against short circuits and slow overcurrents (see figure 20-2). The slow overcurrent element allows the fuse to carry excessive current up to 500% or five times the fuse rating for up to 10 seconds. This will allow the motor to draw the LRA current at start-up for a few seconds without "blowing" the fuse.

The second element provides protection against short circuit currents. This means the dual element fuse can be sized to protect the motor and the circuit wiring, yet allow the motor to draw the excessive starting current. Although the fuse is destroyed and must be replaced, the small cost for replacement will be far less than replacing the motor.

Use the following procedure to select the proper size dual-element fuse for your air conditioning and refrigeration system.

1.　Record the full load amperage rating from the compressor data plate. (In the case of an operating system use the unit's FLA rating.)

2.　Article 440 of the National Electric Code (NEC) states that fuse size must be a minimum of 115% and a maximum of 225% of the compressor FLA rating.

　　a.　Multiply the FLA rating by 1.15 to determine the smallest fuse allowed_____.

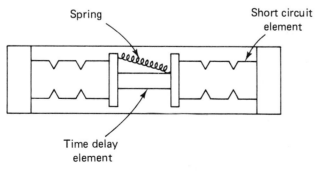

Figure 20-2　Dual element fuse

b. Multiply the FLA rating by 2.25 to determine the largest fuse allowed_____.

   c. The general rule is to install the smallest fuse as determined in step a. A larger fuse is recommended only if the compressor will not start on the smallest fuse.

CIRCUIT BREAKERS

Circuit breakers (see figure 20-3) provide overcurrent protection similar to fuses. The circuit breaker has a bimetal element that moves when heated by excessive current. The movement of the bimetal opens the circuit breaker contacts and interrupts the current flow (see figure 20-4).

   The circuit breaker has characteristics similar to the single element fuse and is not generally used for overcurrent protection on motors.

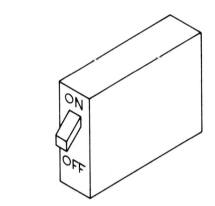

Figure 20-3   Circuit breaker

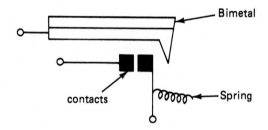

Bimetal moves up as it warms
and unlatches the contacts.
Spring pulls contacts open

Figure 20-4   Circuit breaker diagram

Some circuit breakers are manufacturered with time delay characteristics similar to the dual element fuse and are more expensive than regular circuit breakers.

   The main problem with circuit breakers is that the contacts may sometime weld together during extremely large currents and will not open to provide overcurrent protection. The National Electric Code (section 110-3b)

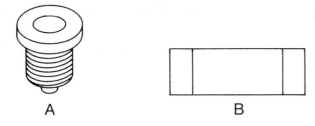

Figure 20–5   (a) Plug type fuse; (b) cartridge type fuse

now requires that all equipment that has the data for maximum fuse size listed on the data plate must use a fuse and not a circuit breaker for overcurrent protection.

TYPES OF FUSES

Fuses are manufactured in several types for different applications. The two most common types are the plug fuse and the cartridge fuse (see figure 20-5).

The plug fuse is intended for branch circuit and small component protection and is made in sizes from 3/10 amps through 30 amps.

A special *size limiting feature* is built into the S type plug fuse. This means that the thread size is not the same for all ampacities (amperage capacity) of the fuses. This prevents a customer from increasing the fuse size if a fuse repeatedly opens. For instance, if a 10 amp fuse is installed to protect a motor and it blows out, the customer cannot replace the 10 amp fuse with a 15 amp fuse.

Cartridge fuses have the same size limiting feature (called a rejection feature). The NEC specifies that all new fuse installations use S type fuses or cartridge fuses with rejection features.

INTERNAL AND EXTERNAL OVERLOADS

The hermetic compressor is cooled by the droplets of refrigerant that are returned through the refrigerant cycle. These droplets are called saturated vapor. The hermetic compressor must have protection against overheating if there is insufficient saturated vapor to cool the motor.

The internal overload and external overload are bimetal-type protection devices that open when they receive too much heat. The internal overload (see figure 20-6) could receive the excessive heat either from overcurrent or the loss of saturated vapor, which is usually due to a refrigerant leak. The internal overload is located in the motor winding. The external overload is located on the dome of the compressor. Since the internal overload is inside the compressor, the complete compressor must be replaced if the internal overload will not reset. The external overload can easily be replaced since it is outside the compressor.

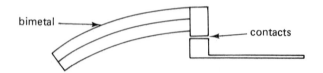

Figure 20-6 Internal overload

When the temperature reaches a point where damage could occur, the internal overload opens and current through the compressor motor winding is interrupted, causing the compressor to stop.

The internal overload may require 12 to 18 hours to cool down and return to the closed position. For this reason you should try to correct the overheating problem, such as loss of refrigerant, before continuing to test the motor.

Your instructor will demonstrate the internal overload protection operation. Follow the steps in this procedure to show the "tripping" or opening of the internal overload and its resetting.

1. *All work in this procedure is to be completed by the instructor as the compressor can easily be damaged by this process.*

2. Turn off all power and connect the following circuit.

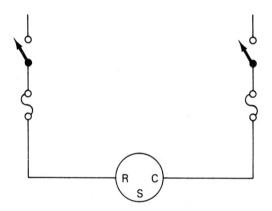

   a. Connect a wire from L_1 to terminal R.

   b. Connect a wire from L_2 to terminal C.

   c. Notice there are no wires connected to terminal S of the compressor. This will cause the motor to draw excessive current in the run winding. The motor will "hum" but not start.

3. Place your clamp-on ammeter around the wire to terminal R and set the range to 100 amps.

4. Apply power and notice the LRA current. The motor winding is now overheating. Allow the overcurrent to exist until the internal overload trips and opens (usually 3 to 12 seconds).

5. After the internal overload opens the motor will stop "humming" and current will be zero.

6. Turn off the disconnect.

7. Set your ohmmeter to R × 1 and zero.

8. Place the ohmmeter probes on terminal R and C of the compressor. Notice there is no continuity and the meter reading is infinity (∞). Leave the meter in place until the overload closes. When the overload closes, the meter will return to a 5 to 12 ohm reading. Note that the time for the internal overload to close will depend on how hot the winding has become.

Service Tip: Notice that the compressor appears to have an open winding. Most technicians would assume the compressor has gone bad and must be replaced. Remember, the internal overload may require up to 18 hours to reset. For this reason always allow a compressor with an open winding to set over night and retest it after it has cooled. This will assure you that the internal overload has had sufficient time to reset.

9. When the internal overload resets, add a run capacitor to the start making the compressor a permanent split capacitor motor.

10. Test run the motor.

11. Notice the overload will not trip if current is within limits.

12. Turn off all power and return all wires to their proper location.

SIZING THE FUSE DISCONNECT, CIRCUIT WIRING AND FUSES

The air conditioning and refrigeration technician must be able to determine the proper size for the circuit wiring, fuse disconnect, and fuse.

Use the following procedure to select the proper size of the fuse disconnect, circuit wiring and fuse.

1. The fuse size must be a minimum of 115% and a maximum of 225% of the FLA rating on the data plate.

2. The size of the wire providing power to the unit must be equal to or larger than the fuse size.

3. The fuse disconnect comes in three basic sizes: 30 amps, 60 amps, and 100 amps. The fuse disconnect must be equal to or larger than the fuse size.

Your instructor will provide several operating systems to check the fuse size, wire size, and fuse disconnect size.

Be sure the power is off before opening the fuse disconnect and checking the wire size and fuse size. Use a current National Electric Code to determine the ampacity of a wire from its gauge size.

Use the following form to record all information.

	Unit 1	Unit 2	Unit 3
FLA rating from data plate	_____	_____	_____
Calculate minimum fuse (FLA × 1.15)	_____	_____	_____
Maximum fuse (FLA × 2.25)	_____	_____	_____
Size of fuse installed	_____	_____	_____
Calculated wire size	_____	_____	_____
Actual wire size installed	_____	_____	_____
Fuse disconnect size	_____	_____	_____

Have your instructor check your information.
Instructor's approval_____

This completes Unit 20. Ask for the quiz on Unit 20 when you are ready.

21 Heating Systems

Tools Required

Voltmeter (250 volt scale)
Ohmmeter (R $\times$ 1K scale)
Screwdriver ($\frac{1}{4}$ in. flat blade)

Objectives for Unit 21

At the conclusion of this unit each student will be able:

To explain orally or in writing the sequential operation of a gas heating system to the instructor's satisfaction.

To explain orally or in writing the sequential operation of an electric heating system to the instructor's satisfaction.

To troubleshoot and repair the electrical components of a gas heating system properly.

To troubleshoot and repair the electrical components of an electric heating system properly.

HEATING SYSTEMS

The air conditioning and refrigeration technician must understand how components operate individually and together as systems. This unit will explain the operation of the heating circuit components as they work together to make a heating system. Typical wiring diagrams will show the relationships between the systems' electrical loads (fan motor and gas valve) and electrical controls (fan switch, limit switch, and thermostat).

BASIC GAS FURNACE OPERATION

The heating system can be directly fueled with gas, oil, or electricity, or be an indirect system such as a heat pump. The simplest system is the gas furnace (see figure 21-1). From the diagram you can see the gas valve. It will be powered with 24 volts when the heating thermostat closes (R to W). At this point the gas valve will energize and allow gas to flow. The gas is ignited by a pilot light and begins to heat the heat exchanger. When the heat exchanger's temperature reaches 120°F, the fan switch will close and the fan will start.

As the room heats up, the thermostat will be satisfied and open R to W, and the gas valve will be de-energized and will close. As the heat exchanger

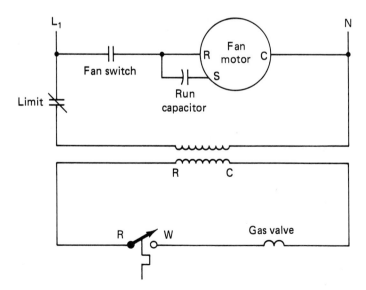

Figure 21-1   Circuit diagram of the basic gas furnace

cools down to 90°, the fan switch will open and turn off the furnace fan, completing the heating cycle.

TESTING THE HEATING SYSTEM

Use the diagram in figure 21-2 to connect the heating components into a heating system which uses natural gas.

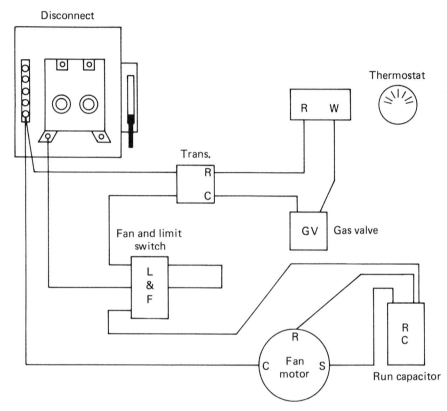

Figure 21-2   Wiring diagram for a gas furnace

Use the following procedure to wire the components and test them. Check off the steps as you complete them.

1. Turn off all power to your trainer.

2. Use figure 21-2 to connect the heating components into the circuit.

3. Ask your instructor to check your wiring before proceeding.
   Instructor's approval _____

4. Turn power on and set the thermostat to the highest number.

5. Set your voltmeter to 50 volts scale and check the gas valve to see if it is receiving 24 volts.
   a. Place one meter probe on each terminal of the gas valve (this is called *measuring across* the gas valve).
   b. If you do not have 24 volts, call your instructor to check your circuit.

6. Since you do not have a fire, use a lamp or heat gun to apply heat to your fan switch.

   a. The fan should turn on after a few minutes of heating.

   b. Test the fan motor for voltage across terminal R to C.

   c. Notice that if the motor is energized you will read 115 volts across R to C.

   d. If the meter does not read 115 volts the fan switch has not closed. Continue heating the fan switch until it closes.

7. Once the fan has started to operate, measure and record its amperage _____. Check the amperage reading against the motor's data.

8. Apply heat to the thermostat to bring the room temperature up to set point.

   a. What happens when the thermostat is satisfied? _____
   _____
   _____

   b. Measure the voltage across the gas valve after the thermostat has opened. What is the voltage across the gas valve now?_____
   _____
   _____

   c. Allow the fan switch to cool also. When the fan switch opens, measure the voltage across the fan motor R to C. What is the fan voltage? _____

9. Repeat steps 4 and 6.

   a. Allow the heat to remain on the limit switch until it opens. (Remember: the fan switch and limit switch have the same bimetal operator.)

   b. Measure and record the voltage across the gas valve after the limit switch has opened. _____

   c. Explain what has happened to the furnance._____
   _____
   _____

10. Turn off power to your trainer and leave the wiring. Call your instructor to discuss the total furnace operation. Be sure you fully understand the operation before you proceed to troubleshoot the system.

TROUBLESHOOTING THE HEATING SYSTEM

The air conditioning technician must be able to observe a malfunctioning heating system and determine which component is causing the problem.

The technician must then use the proper test procedure to determine the condition of parts that are suspected to be faulty. Use the following procedure to troubleshoot the gas heating system. Check off the steps as you complete them.

1. Ask your instructor to put a problem in your heating system.
2. Try to make your system run.
   a. Turn on power.
   b. Set thermostat to the highest setting.
   c. Check gas valve to see if it is energized.
   d. Add heat to the fan switch and check the fan motor operation.
3. At this point make a short appraisal of the conditon of the system. Pay particular attention to parts that are operating correctly and parts that are not._____
   _____
   _____
   _____
   _____
   _____
   _____

Use the following diagram (figure 21-3) for troubleshooting.

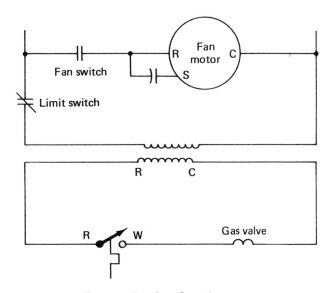

Figure 21-3   Gas furnace

4. If none of the parts operate, check the voltage and record_____
   _____. If the voltage is zero, check the fuses and power source.
5. Check voltage across the gas valve. If no voltage is present, skip to step 12.
6. Since there is supply voltage but no voltage to the gas valve, continue to step 7.

7. Check the transformer voltage at R to C.

    a. If voltage is zero replace the transformer and return to step 1.

    b. If voltage is present skip to step c.

8. Leave one probe on transformer terminal C for steps 9 to 11.

9. Move the other probe to terminal R of the thermostat.

    a. If voltage is zero, the wire between the transformer and terminal R on the thermostat is bad. Replace it and return to step 1.

    b. If voltage is 24 volts, continue to step 10.

10. Move the probe to terminal W on the thermostat.

    a. If voltage is zero, the thermostat is bad. Replace the thermostat and return to step 1.

    b. If voltage is 24 volts, continue to step 11.

11. Move the probe to the gas valve terminal where the wire from terminal W on the thermostat connects to the gas valve.

    a. If voltage is zero, the wire between the thermostat and gas valve is bad. Replace the wire and return to step 1.

    b. From your diagram you can see if the voltage is 24 volts. The wiring and switches on the left side of the gas valve circuit are good.

    c. Since there is only one wire and no controls on the right side of the gas valve, the problem must be in the wire between transformer terminal C and the gas valve. Replace this wire and return to step 1.

12. Since 24 volts is present across the gas valve, the valve must be inoperative. Check the valve with an ohmmeter to confirm that it is bad. Replace the valve and return to step 1.

13. Call your instructor at this time to discuss the method used in steps 1 to 12 to find a problem. Be sure you can explain the following:

    a. Why do you try to run the system first?

    b. Why do you check for voltage across the gas valve before looking for bad wires?

    c. What is the condition of the controls and the wires if you measure zero volts across the gas valve, but have 24 volts across the transformer R to C?

    d. What is the condition of the controls and the wiring for the gas valve if you read 24 volts across the gas valve?

    e. Why must you leave one probe at terminal C to test the wires on the left side of the circuit?

    f. Where would you leave the probe if you wanted to test the wires on the right side of the circuit?

Your instructor will place a problem in the fan circuit after you have *mastered* steps 1 to 12. Use the same method to find problems in the fan circuit that you used in steps 1 to 12. Remember: this is the method to use every time you must troubleshoot a furnace.

BASIC ELECTRIC HEATING SYSTEM OPERATION

The electric furnance uses current flowing through high resistance heating coils to produce heat. From the diagram in figure 21-4 you can see that the system uses a low voltage thermostat, terminals R to W, to control voltage to the sequencer's coil at terminals H_1 and H_2. The sequencer coil heats the bimetal contacts at M_1 and M_2 in the sequencer. When these contacts close, current flows to the furnace heating coils. A second set of auxiliary contacts (A_1 and A_2) will close at the same time as the main contacts. The auxiliary contacts control current to the fan relay coil. When the fan relay's coil is energized, it will close the fan relay contacts and turn on the furnance fan.

When the room has warmed to the set point temperature, the thermostat opens the circuit R to W, stopping current to the sequencer's coil. The bimetal cools down slowly, giving a one to two minute time delay before opening the main and auxiliary contacts. After the heater has cooled down, the operation is ready to start again when the room cools below the thermostat set point.

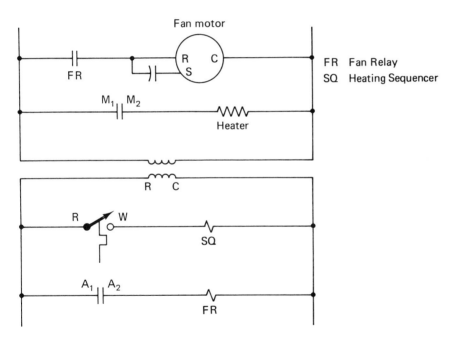

Figure 21-4   Typical diagram of an electric furnace

WIRING THE ELECTRIC FURNACE

Use the diagram in figure 21-4 and the following procedure to wire the electric heating components on your trainer. Check off the steps as you complete them. Test run the system *after* your instructor has checked your wiring.

1. Turn off the power to your trainer.

2. Connect the wires of the low voltage circuit first.

   a. Connect a wire from transformer terminal R to thermostat terminal R.

   b. Connect a wire from thermostat terminal W to the sequencer coil terminal H_1.

   c. Connect a wire from sequencer coil terminal H_2 to terminal C on the transformer.

   d. Connect a wire from transformer terminal R to the sequencer auxiliary contact A_1.

   e. Connect a wire from auxiliary contact A_2 to one of the coil terminals on the fan relay.

   f. Connect a wire from the other fan relay coil terminal to terminal C on the transformer.

3. Now connect the high voltage circuit wires.

   a. Connect the furnace fan as a PSC motor.

   b. Connect the fan motor in series with the fan relay contacts and L_2.

   c. Connect a wire from L_1 to the other fan relay contact terminal.

   d. Connect a wire from L_1 to terminal M_1 on the sequencer.

   e. Connect the electric heating coil (lamp) between terminal M_2 on the sequencer and L_2.

4. Ask your instructor to inspect your wiring. Instructor's approval_____
   _____.

5. Apply power and check the operation of your system. Leave the circuit in place and proceed to the section on troubleshooting.

TROUBLESHOOTING THE ELECTRIC HEATING SYSTEM

Refer to figure 21-4 to help you troubleshoot your electric heating system. Use the following procedure and check off the steps as you complete them. Ask your instructor to place a problem in your heating circuit at this time.

1. Set the thermostat to the highest number in heating. Apply power and try to make the system operate.

2. Observe the system to determine what components are operating correctly. Be sure to allow one to two minutes for the sequencer's time delay to operate.

   a. If none of the components are operating, skip to step 3.

   b. If one component is operating and others are not, skip to step 6.

   c. If the circuit is operating correctly, turn the power off and ask your instructor for a new problem.

3. Since none of the components are operating, make the following checks to determine if the system has proper voltage.

4. Check power at L_1 and L_2 in the disconnect.
   a. If voltage is not present, check the fuses and return to step 2.
   b. If voltage is present at the disconnect, check voltage at terminals R to C on the transformer.
   c. If voltage is not present at R to C, change the transformer and return to step 2.

5. If voltage is present at R to C, check for voltage across terminals H_1 to H_2 on the sequencer coil.
   a. If no voltage is present at H_1 to H_2, use the voltage drop method to find the point where voltage has stopped in the circuit from the thermostat to the sequencer coil. Make repairs and return to step 2.
   b. If voltage is present at H_1 and H_2, check for voltage across the furnace heater coil (lamp).
   c. If no voltage is present at the furnace heating coil, use the voltage drop method to locate where the voltage has stopped. Make repairs if necessary and return to step 2.
   d. Check the fan circuit for proper voltage if necessary.

6. If the fan or furnace heating coils are operating, you may assume that the voltage to the transformer and controls is correct. Concentrate your checks on the circuit of the component that will not operate.
   a. Use the voltage drop method to determine where voltage has stopped.
   b. If you have applied voltage across the fan motor terminals R to C and S to C during start and the motor fails to run, test the fan motor for an open overload and make repairs as necessary and return to step 2.
   c. If you have applied voltage across the furnace heating coils and it fails to heat or draw current, test the heating coils for continuity (with the power off). Replace the coil if necessary and return to step 2.

After you have successfully found each problem, turn off the power to your trainer and remove all wiring.

This completes Unit 21. Ask your instructor for the quiz on Unit 21 when you are ready.

22 The Refrigeration System

Tools Required

Volmeter (250 volt scale)

Clamp-on ammeter (30 amp scale)

Screwdriver (¼ in. flat blade)

Objectives for Unit 22

At conclusion of this unit each student will be able:

To explain orally or in writing the sequential operation of a refrigeration system to the instructor's satisfaction.

To explain orally or in writing the sequential operation of the electric-heater-type defrost cycle to the instructor's satisfaction.

To explain orally or in writing the sequential operation of the hot-gas type defrost cycle to the instructor's satisfaction.

To wire a refrigeration system correctly and check the components for normal operation.

To demonstrate the correct use of the voltage drop troubleshooting method to locate a faulty component or wire.

To find faulty components or wires in the refrigeration system.

THE REFRIGERATION SYSTEM

The individual parts of the refrigeration system have been discussed in previous units. In this unit the components will be wired together to simulate an operating refrigeration system. Emphasis will be placed on the operation and troubleshooting of this system.

OPERATION OF THE REFRIGERATION SYSTEM

The diagram in figure 22-1 shows the electrical components used in a commercial refrigeration system. Most small commercial systems such as display cases or reach-in freezers use line voltage controls instead of a relay. This means that the pressure controls, temperature controls and defrost clock will be wired in series with the compressor and must be rated large enough to carry the compressor current.

From the diagram you can see that when the thermostat closes the compressor, evaporator fan and condenser fan will start running and stay on until the conditioned space is cooled to the predetermined set point. The thermostat will open at the set point temperature and turn off these components. When the conditioned space warms up two to three degrees, the thermostat will close and start the cycle again.

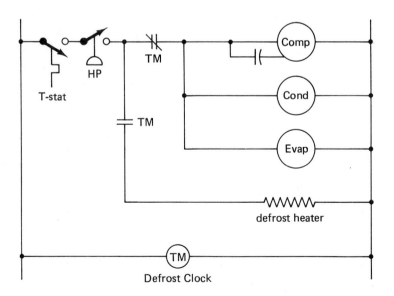

Figure 22-1   Refrigeration system with defrost heater

After the system has run for 12 hours, the defrost clock will turn on the defrost cycle. The defrost cycle in figure 22-1 uses an electric heater that is placed on the evaporator coil to melt the frost from the fins. The compressor, condenser fan, and evaporator are all turned off during the defrost cycle. When the defrost cycle terminates, the electric heater is turned off and the compressor, condenser fan, and evaporator fan are again energized.

Another defrost method is called the hot gas defrost. In this method the compressor and condenser fan stay on and a valve opens to allow hot refrigerant gas from the compressor to go directly to the evaporator to melt off the frost. In figure 22-2 you can see that the defrost cycle is still controlled by a defrost clock.

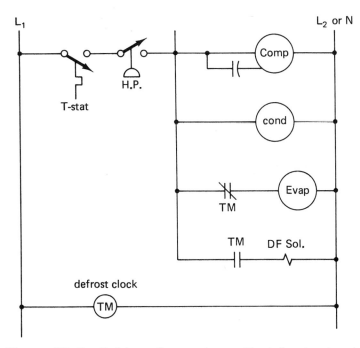

Figure 22-2   Refrigeration system with defrost solenoid

The normally *open* contacts in the defrost timer close during defrost and energize the defrost solenoid valve. The normally *closed* contacts open and turn off the evaporator fan. After the defrost cycle has ended the system is reversed: the evaporator fan is turned on and the defrost solenoid is turned off.

Use the following procedure to wire the components of a refrigeration system and test run them. Check off the steps as you complete them.

1. Turn off all power to your trainer.

2. Use the diagram in figure 22-3 to complete the wiring of your system.

3. Ask your instructor to approve your wiring.

   Instructor's approval _____

4. If you used a low pressure switch for a temperature control, apply refrigerant pressure to close the switch.

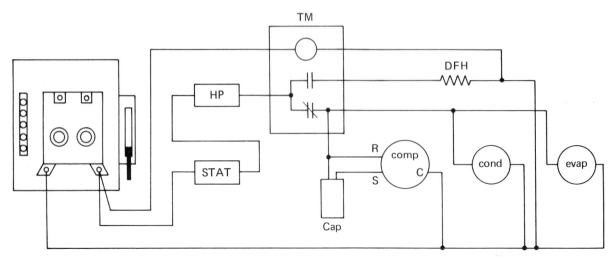

Figure 22-3   Refrigeration system with defrost heater

5.   If your system uses a thermostat for temperature control, set the thermostat to the lowest temperature.

6.   Set your ammeter on the 30 amp scale and place it around the wire connected to the R terminal on the compressor. Be sure to observe the starting and safety precautions you learned about compressors.

7.   Apply power to the circuit and observe the compressor, condenser fan and evaporator fan to be sure they are all operating. NOTE: If any of the motors are not operating, turn the power off *immediately* and call your instructor. Your instructor may find the problem for you or direct you to the troubleshooting procedure in this unit.

8.   Adjust the temperature control (or pressure control) to a setting to turn off the components. Be sure the compressor, condenser fan, and evaporator fan all turned off. If any of these motors stay on, see the note in step 7.

9.   Turn the temperature control on again. Be sure all the motors are on again.

10. Adjust the high pressure safety control so it opens and turns off the compressor, evaporator fan, and condenser fan. (Refrigerant pressure will need to be added to the high pressure control to cause it to open.) If all the motors do not stop, refer to the note in step 7.

11. Allow the high pressure to return to normal and turn on the motors again.

12. Advance the defrost timer until the defrost cycle begins.

   a.   Be sure the compressor, evaporator fan, and condenser fan all stop.

   b.   Use your clamp-on ammeter to check the defrost heater for correct operation. If there is no current flow, refer to the note in step 7.

13. Advance the defrost clock to terminate the defrost cycle.

a. The defrost heater should go off.

b. The compressor, condenser fan, and evaporator fan should go on.

14. Turn off power to the system and change the defrost method as shown in the diagram in figure 22-4.

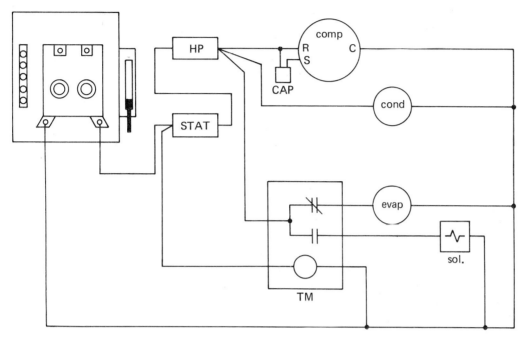

Figure 22–4   Refrigeration system with defrost solenoid

15. Turn the power on and set the temperature control so the compressor, condenser fan, and evaporator fan are all operating.

16. Advance the defrost clock to a defrost cycle.

a. This time only the evaporator fan should go off. The compressor and condenser fan should stay on.

b. The defrost solenoid should "click on."

Note: If you are having trouble determining if your solenoid is off or on, check for applied voltage across its terminals. If applied voltage is present, the solenoid is on.

17. Advance the defrost timer to end the defrost cycle.

a. The evaporator should come on again.

b. The defrost solenoid should go off.

18. Turn off the power and leave all wiring in place.

TROUBLESHOOTING THE REFRIGERATION SYSTEM

The following procedure will help you determine the condition of the components in the refrigeration system. Ask your instructor to place a problem in your system at this time. You will need to refer to the operation steps at times to try to operate the system. Check off the steps as you complete them.

1. Try to make the system run. Use ammeter to determine if all the motors are operating correctly.

   a. If all motors are not operating, continue to step 2.

   b. If all motors are operating correctly, check the defrost cycle and refer to step 4.

2. Use the ammeter to help you determine which motors are not operating correctly. You can visually check the fan motors, but you should also check for proper amperage to the motors.

3. Check the inoperative motor with a voltmeter.

   a. If correct voltage is present across the motor terminals but the motor will not operate, the motor is bad. Replace the motor and return to step 7.

   b. If no voltage is present across the motor terminals, use the *voltage drop* method to locate the voltage loss. Figure 22-5 will help you with these checks. This example uses the compressor, but can also be used for the evaporator or condenser fan.

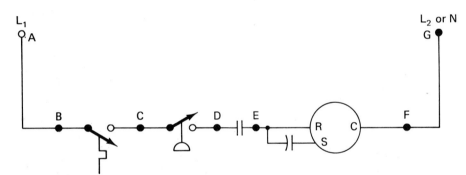

Figure 22-5   Voltage drop test points

   c. Test the voltage A to G. If no voltage is present, check the supply fuses and replace if necessary. Then return to step 1.

   d. If there is voltage between points A and G but not across the run and common terminals of the motor, your circuit has an open control or bad wire.

   e. Leave one probe on terminal G and test the following points in order and record the voltage reading.

   B _____

   C _____

   D _____

   E _____

The point where voltage is zero indicates the problem is between that point and the point prior to it. (For example, if you measured

230 volts at point C but zero at Point D, it shows that the pressure switch is open.)

If you measured applied voltage at point E, you have demonstrated that all the controls and wires from point A to E are all operating correctly. This means the problem must be in the wire connecting terminal C on the compressor to L_2 (point G).

    f.   Leave one probe on point A and move the other probe to point F. If no voltage is measured it shows that the wire between point G and F is bad. Replace it and return to step 1.

It is important to remember this procedure; the voltage drop check for a bad wire or open component is the *simplest and most reliable method to find the problem.* If you do not understand this method, stop and ask your instructor for help. DO NOT PROCEED IF YOU CAN NOT CORRECTLY MAKE THE VOLTAGE DROP TEST for finding a bad wire or component.

4.   Since all motors in steps 1 to 3 were working correctly, advance the defrost control and check the defrost cycle.

    a.   If the system does not operate correctly, use the voltage drop method to find the component or wire that is faulty. Make the required repairs and return to step 1.

    b.   If the system is operating correctly shut off all power and ask your instructor to place a new problem in the system. When you have found all problems, shut off all power and unwire the system.

Your instructor may take you to observe an operating refrigeration system. Remember that it uses components and controls similar to those you have used on your trainer.

This completes Unit 22. Ask your instructor for the quiz on Unit 22 when you are ready.

23 Air Conditioning Systems and Heat Pumps

Safety for Unit 23

At times the circuits and disconnects you will be working with will be powered with up to 230 volts. You must be aware of and take appropriate safety precautions relative to electrical shock hazards and wear safety glasses while working.

You will also be working with open-type motors. Safety precautions relative to open rotating shafts must be observed. If your instructor takes you to observe an operating air conditioning system, you will be given a list of safety rules pertaining to operating equipment. Please observe these precautions when observing equipment (or operating equipment).

Tools Required

Voltmeter (250 volt scale)

Clamp-on ammeter (30 amp scale)

Screwdriver (¼ in. flat blade)

Objectives for Unit 23

At the conclusion of this unit each student will be able:

To explain orally or in writing the sequential operation of an air conditioning system to the instructor's satisfaction.

To wire an air conditioning system correctly and check the components for normal operation.

To find faulty components or wires in the air conditioning system.

To explain orally or in writing the sequential operation of a heat pump in the cooling, heating, and defrost modes.

To wire a heat pump correctly and check the components for normal operation.

AIR CONDITIONING SYSTEMS AND HEAT PUMPS

The individual parts of the air conditioning system have been discussed in previous units. In this unit the components will be wired together to simulate an operating air conditioning system. Emphasis will be placed on the operation and troubleshooting of this system. .

OPERATION OF THE AIR CONDITIONING SYSTEM

The diagram in figure 23-1 shows the components of a basic air conditioning system.

The system becomes energized when the room thermostat closes. This occurs when the temperature in the conditioned space gets warmer than the

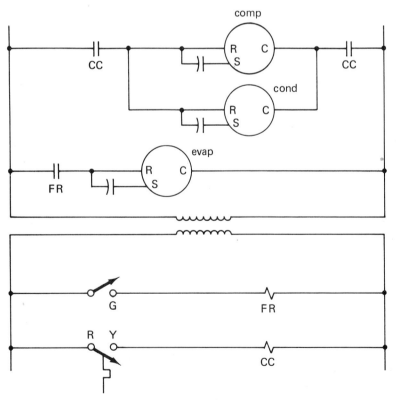

Figure 23-1   Air conditioning system

thermostat set point temperature. The thermostat closes the circuit to the coil of the compressor contactor. As current flows through the contactor's coil it becomes magnetized and pulls the compressor contactor's normally open contacts closed. When these contacts close, current flows to the compressor. The compressor is energized and begins to pump refrigerant. The condenser fan is connected in parallel with the compressor and is energized at the same time. This means the condenser fan should be on when the compressor is running and off when the compressor is not running.

The thermostat terminal Y will provide power to terminal G through the thermostat fan switch when it is set in the AUTO position. Terminal G energizes the fan relay coil. When the fan relay coil is magnetized, it pulls the fan relay contacts closed and the evaporator (furnace fan) begins to operate. This means the evaporator fan will come on at the same time the compressor and condensor fan comes on when the fan switch is set in the AUTO position. If the fan switch is set in the ON position, terminal G will be powered at all times and the evaporator fan will run continuously, even when the compressor and the condenser fan are turned off.

When the room temperature cools below the thermostat set point temperature, the thermostat contacts between R and Y will open and the compressor contactor's coil will be de-energized. At this point the compressor contactor's contacts will open and turn off the compressor and condenser fan. If the fan switch is in the AUTO position, the fan relay's coil will be de-energized when the R to Y contacts open and cause the fan relay's contacts to open. This will turn off the evaporator fan. If the fan switch is in the ON position, the fan relay coil will continue to be energized and the evaporator fan will stay on.

TEST RUNNING THE AIR CONDITIONING SYSTEM

Use the diagram in figure 23-2 to connect the air conditioning components on your trainer. Use the following procedure to test run the air conditioning system. Check off the steps as you complete them.

1. Turn off all power to your trainer.

2. Connect the low voltage circuit of the compressor contactor's coil.

3. Connect the low voltage circuit of the fan relay's coil.

4. Connect the furnace fan through the fan relay contacts. Wire the fan motor as a PSC motor.

5. Connect the compressor through the compressor contactor's contacts. Wire the compressor motor as a PSC motor.

6. Connect the condenser fan in parallel with the compressor. Wire the condenser fan as a PSC motor.

7. Ask your instructor to inspect your wiring *before* you apply power.
   Instructor's approval _____

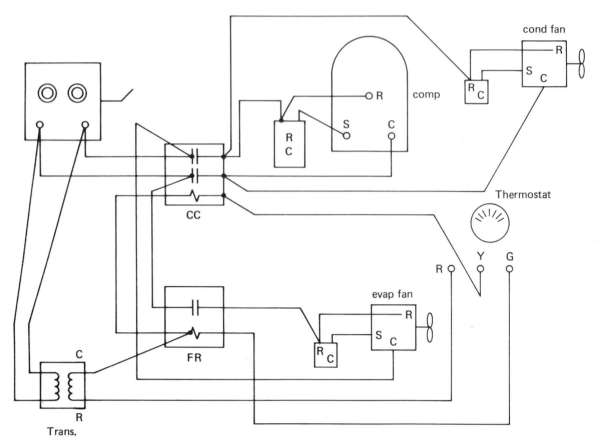

Figure 23–2 Air conditioning system

8. Set the thermostat to the lowest setting in cooling and the fan switch to AUTO.

9. Apply power and use the clamp-on ammeter to be sure each motor is operating correctly. If any of the motors fail to start, turn the power off immediately and ask your instructor to recheck the system.

10. Adjust the thermostat to the warmest setting. Notice that all the motors turn off.

11. Set the fan switch to ON. The fan motor should turn on.

   a. Set the thermostat at the lowest setting to turn on the compressor and condenser fan.

   b. After the compressor has run for several minutes, set the thermostat to the highest setting and notice that the compressor and condenser fan turn off and the evaporator (furnace) fan stays on.

   c. Explain why the home owner may wish to run the furnance fan by itself. _____

      _____

      _____

      _____

d. Explain why the furnace fan must be on any time the compressor is running. _____

_____

_____

Have your instructor check your answers.

12. Leave the circuit connected for the troubleshooting procedure.

TROUBLESHOOTING THE AIR CONDITIONING SYSTEM

As an air conditioning technician you will be sent to troubleshoot inoperative air conditioning systems. You must be able to identify the problems, test the components, and make repairs as quickly as possible. You must understand that approximately 80% of all problems in air conditioning systems involve electrical systems.

Use the following procedure to test the air conditioning system on your trainer. This procedure can also be used to test operating systems you will encounter in the field.

Use the diagrams in figures 23-1 and 23-2 to help you to find the problems in this circuit. Check off the steps as you complete them. Ask your instructor to put a problem in your system.

1. Set the thermostat to the lowest setting. Turn on the disconnect and try to make the system operate.

    a. If none of the motors start, skip to step 2.

    b. If one or two of the motors start, but one or more motors do not run, skip to step 3.

    c. If all the motors operate correctly and pull the correct FLA, turn the power off and ask your instructor for a new problem.

2. Since no motors are operating, make these tests for voltage.

    a. L_1 and L_2 at the disconnect; if voltage is not present check the fuses and return to step 1.

    b. If voltage is present at L_1 to L_2, check voltage at R to C on the transformer.

    c. If voltage is not present at R to C on the transformer replace the transformer and return to step 1.

    d. If voltage is present at R to C, check for voltage across the compressor contactor's coil.

    e. If 24 volts is not present at the contactor's coil, the problem is in the thermostat or circuit. Use the voltage drop method to locate the problem. Make repairs as necessary and return to step 1.

3. Since one or more of the motors are operating, you may assume that the voltages at the disconnect and the transformer secondary are correct. (Hint: the motors can only start when the 24 volt relay is energized.)

a. Identify the motor that is not operating. Use the diagrams in figures 23-1 and 23-2 to determine the controls that turn the motor on and off.

4. Let's use the evaporator fan circuit as an example. Since the evaporator fan does not operate, you will identify the fan relay contacts as the control that turns the fan motor off and on. The fan relay coil controls the fan relay contacts, so you must determine if the fan relay coil is energized. Use the diagram in figure 23-3 to help you make your checks.

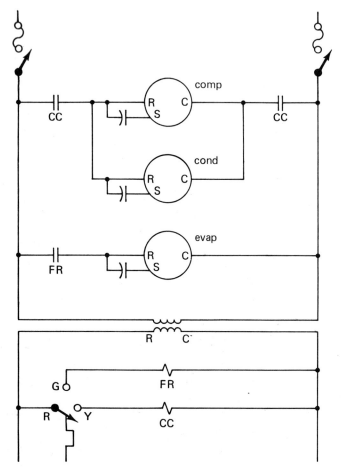

Figure 23-3

a. Check for 24 volts across the fan relay coil.

b. If 24 volts is present, check to see if the relay contacts have closed. *SERVICE TIP:* Move the fan switch from AUTO to ON with the thermostat set to the highest temperature. This will energize and de-energize the fan relay coil, and you should hear the contacts "click"; replace the fan relay and return to step 1.

c. If the relay "clicks," check across R to C and S to C to be sure there is applied voltage. If applied voltage is present and the motor will not start, the problem is in the motor. Check the motor's temperature to determine if the overload is open. If the motor is cool, replace the motor and return to step 1.

d. If no voltage is present at the motor, the problem is voltage loss in the circuit, and you will need to use the voltage drop troubleshooting method to locate the problem. Leave one probe of your voltmeter to L_2 and check the following points for voltage.

1. L_1

2. Line side of the fan relay contacts

3. Load (motor) side of fan relay contacts

4. Terminal R on the motor

5. Red dot terminal on the run capacitor

6. Terminal S on the motor

The point between where voltage is last measured and where voltage is lost is the fault or the broken part of the circuit. Make necessary repairs and return to step 1.

If you measured applied voltage at each of these test points, your problem is the wire connecting terminal C to L_2. Remove the wire and check for continuity. Replace the parts as needed and return to step 1.

If you have found the problem easily, ask your instructor to place another problem in your circuit. It is very important to remember there is not a "best" method to find a problem in the circuit. You will be required to use a variety of these methods: visual inspection, voltage measurements, current measurements, continuity measurements, and the voltage drop trouble-shooting method. After you have found all problems in the system, shut the power off and leave your circuit in place.

OPERATION OF THE HEAT PUMP

The heat pump is very similar to the air conditioning system in that it has a compressor, condenser fan, and evaporator fan. It is different from the air conditioning system in that it can operate to cool or heat. In the heating mode the refrigeration flow is reversed by a reversing valve. The reversing valve causes the hot refrigerant vapor from the compressor to be routined to the indoor coil (in the furnace) where it is condensed and heat is removed. The indoor fan (furnace fan) moves this heat through the house to warm it. The condensed refrigerant moves to the outdoor coil where it evaporates and picks up more heat. In the heating cycle the outdoor coil is the evaporator and the indoor coil is the condenser.

The outdoor coil can remove heat from the air down to below 0°F. Many manufacturers use controls to turn off the unit at approximately 0°F for economical reasons. At this time supplemental heat is turned on. This heat usually is supplied by electrical resistance coils similar to those in the electric furnace.

When outdoor temperatures are below 40°F frost will begin to form on the outdoor coil. The frost is removed about every 90 minutes by reversing the refrigeration cycle back to the cooling mode. This routes the hot refrigerant gas to the outdoor unit. The hot gas melts the frost on the outdoor coil. Since the system is in the cooling mode during defrost, the indoor coil is the evaporator and cool air is being moved by the furnance fan. To temper the indoor air during defrost the electric heat is turned on. A defrost control determines when the defrost cycle will begin and end.

In the cooling mode you can see from figure 23-4 that the cooling thermostat will close, completing the circuit R to Y and energize CC, the compressor contactors coil. The contactor's coil will become magnetized and pull in the compressor contactor's normally open contacts. When the contacts close, the compressor and condenser fan will both start running.

Terminal G will be powered from terminal Y at the thermostat fan switch. Terminal G will energize the fan relay coil (FR) and pull in the fan relay contacts. This will cause the indoor (furnace) fan to start at the same time as the compressor and condenser.

The reversing valve solenoid is de-energized in the cooling mode. This will route the refrigerant to the outdoor coil to be condensed and to the indoor coil to be evaporated. The defrost control will not affect the cooling mode, since frost will not form on the coils when the system is cooling.

In the heating mode, the heating thermostat contacts R to W will close. This will energize the heating relay coil (HR) and close two sets of heating relay contacts. One set of heating relay contacts will close and energize the compressor contactor's coil (CC) causing the CC contacts to close. This will energize the compressor and condenser fan. The second set of heat relay (HR) contacts will energize the reversing valve solenoid. When the reversing valve solenoid is energized, the refrigerant route is changed so that the indoor coil in the furnace becomes the condenser and the outdoor coil becomes the evaporator.

When the coil becomes frosted the defrost control will open its normally closed contacts and de-energize the outdoor fan and reversing valve, and energize the auxiliary heat. When the reversing valve is de-energized the system returns to the cooling mode, and hot gas is directed to the frosted outdoor coil. By turning off the outdoor fan the defrost process is speeded up. The auxiliary heat is needed to temper or warm the indoor air. When the defrost cycle ends (it usually lasts no longer than 10 minutes), the defrost control closes its contacts and energizes the outdoor fan and reversing valve, and de-energizes the auxiliary heat. The system returns to the normal heating cycle until the room temperature is satisfied. The heat pump adds heat to the room very slowly so it may stay on for longer periods of time than a gas or oil fired furnace. Then the room thermostat heating contacts R to W will open and turn the heat pump off.

Many manufacturers of heat pumps add extra circuits for emergency heat, second and third stage heat, and outdoor temperature controls. You can be assured that all systems will operate similarly to the basic heat pump system just covered.

TEST RUNNING THE HEAT PUMP

Use the diagram in figure 23-4 to connect the controls on your trainer to simulate a heat pump system. Use the following procedure to test run the

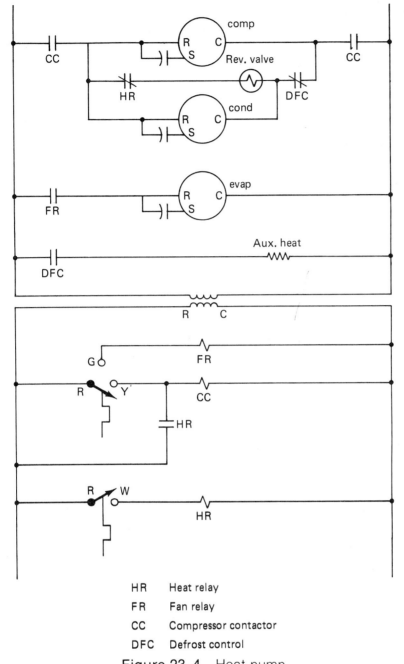

HR	Heat relay
FR	Fan relay
CC	Compressor contactor
DFC	Defrost control

Figure 23-4   Heat pump

system. Check off the steps as you complete them. Use the following components and check for proper voltage.

1. Turn off all power to the trainer.

2. Use the solenoid valve on your trainer to represent the reversing valve.

3. Use a lamp for the electric resistance heat.

4. Use a relay with a set of normally open and normally closed contacts for the defrost control. (Jump the proper voltage to the coil to energize the defrost control as needed.)

5. The thermostat, fan relay, and all motors will be the same as in the air conditioning system.

6. Have your instructor check your circuit wiring and check all components to see that they are connected to the proper voltages.

7. Turn on power and make the system run in the cooling mode.

   _____
   _____
   _____
   _____

8. Turn the system to the heating mode and identify the components that are energized (operating). _____

   _____
   _____
   _____

9. Energize the defrost control (relay) and identify the components that are energized (operating). _____

   _____
   _____
   _____

10. Turn all power off and return all wires and components to their proper place.

    At this time your instructor will take you to observe an operating heat pump. Be sure to observe the safety rules and procedures for operating equipment.
    This completes Unit 23. When you are ready ask your instructor for the quiz on Unit 23.

24 Commercial System Controls

Tools Required

Voltmeter (250 volt scale)

Clamp-on ammeter (30 amp scale)

Screwdriver (¼ in. flat blade)

Objectives for Unit 24

At the conclusion of this unit the student should be able:

To explain orally or in writing the operation of the motor starter and its overloads and heaters to the instructor's satisfaction.

To explain orally or in writing the operation of a start-stop switch to the instructor's satisfaction.

To wire a start-stop switch correctly to a starter and operate the starter.

To find faulty components or wires in the start-stop switch and starter circuit.

COMMERCIAL SYSTEM CONTROLS

Commercial air conditioning and refrigeration systems use the motor starter to control large motors. The basic motor starter pictured in figure 24-1 is very similar to a relay, in that it has a coil that becomes a magnet and pulls a set of contacts closed.

The motor starter is more complex than a relay in that it incorporates an overload in series with its coil.

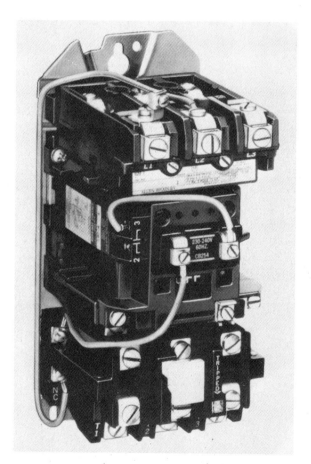

Figure 24-1   Starter (courtesy of *Allen-Bradley Company*)

The overload senses heat from heaters that are in series with the contacts (see figure 24-2). When excessive current flows through the contacts and heaters into the motor, the heater gives off heat that builds to a set point and "trips" or opens the overload. When the overload opens it interrupts current

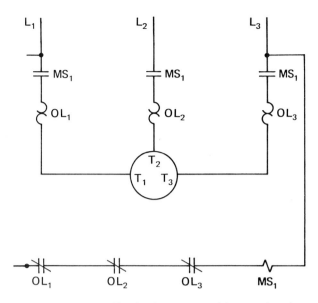

Figure 24-2 Typical starter with overloads

to the starter magnetic coil. The coil loses its magnetism, allowing the contacts to return open. After the overload cools down, the overload on the starter can be reset and the coil can again be energized.

THE START-STOP SWITCH

The starter is generally used with one or more start-stop switches to control refrigeration and air conditioning motors. Figure 24-3 shows the electrical symbols and a photo of the start-stop switch. The start switch is a normally open pushbutton switch. The pushbutton switch will remain in the normal position because of spring tension. When the normally open pushbutton switch is manually depressed, its contacts close, and remain closed only while someone is pushing on the button. As soon as the finger is removed from the button it returns to open.

The stop switch is a normally closed pushbutton switch. Its contacts remain closed until its button is depressed and will return to closed as soon as the finger leaves the switch.

TESTING THE START-STOP SWITCH

Use the following procedure to test the start-stop switch. Check off the steps as you complete them.

1. Leave all power off for this test.

2. Locate the start-stop switch on your trainer and remove its cover.

3. Set your ohmmeter on R X 1.

4. Locate the contacts for the start switch and test them for continuity. Notice that the meter shows they are open.

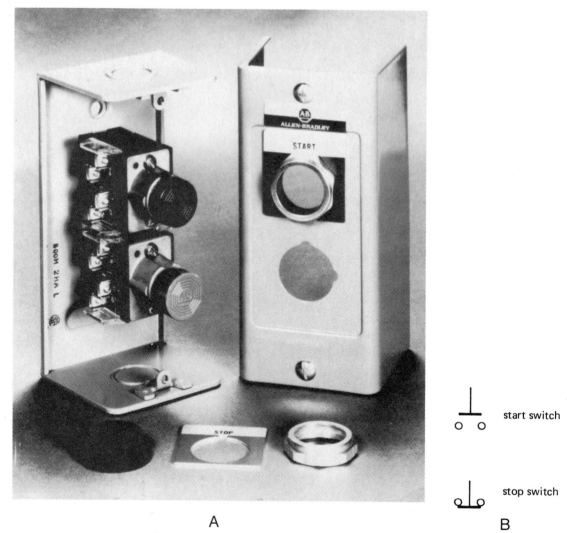

start switch

stop switch

A   B

Figure 24-3 Photograph and symbols of start-stop switch (courtesy of *Allen-Bradley Company*)

5. Push the start button while the ohmmeter is connected across the start switch contacts. Notice that the meter shows they are closed.

6. Locate the contacts for the stop switch and test them for continuity. Notice that the meter also shows they are closed.

7. Push the stop button while the ohmmeter is connected across the stop switch. Notice they open when the stop button is depressed.

8. If the start and stop switch do not behave as outlined above, call your instructor for help.

9. Replace the cover on the start-stop switch.

TYPICAL STARTING CIRCUIT

The diagram in figure 24-4 shows the start-stop switch connected to the starter. Notice that the start and stop switch contacts are in series with the

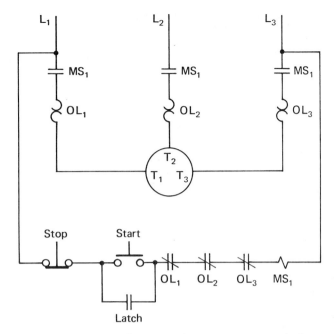

Figure 24-4 Starter with start-stop switch

starter coil. The starter overload is normally closed and in series with the coil. A set of *seal or latch contacts* are in parallel with the start button. If the seal contacts were not added, you would have had to have the start button pushed in as long as you wanted the motor to operate, since the start switch will return to its normally open condition as soon as you remove your finger from it. The seal contacts are activated by the starter coil and will close with the main contacts the instant the start button energizes the starter coil. The seal contacts provide a parallel route around the open start button contacts, and yet allow the stop button to de-energize the coil circuit when required.

The main contacts (usually three sets) will close when the coil becomes magnetized. They provide the main path for current to the motor. If the motor is single phase, only two sets of main contacts will be used.

This circuit is typical of most starter circuits. Some variations may be made by some equipment manufacturers, but the basic concepts remain the same.

WIRING AND TESTING THE STARTER CIRCUIT

Use this procedure to wire the starter in a circuit. Refer to figure 24-5 to locate the starter parts. Check off the steps as you complete them.

1. Turn off all power to your trainer.

2. Locate the starter on your trainer.

3. Inspect the starter and locate the two coil terminals. Identify the coil voltage. NOTE: Most starter coils can be changed for different voltages, usually 120 or 230 volts.

4. Locate the overloads for each of the two heaters.

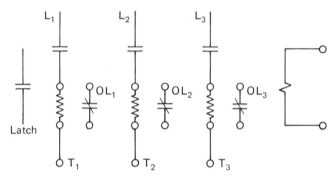

Figure 24-5   Starter

5. Locate the start-stop switch.

6. Connect the stop switch to L_1.

7. Connect the other terminal of the stop switch to the start switch.

8. Connect the other terminal of the start switch to the overload contacts number 1.

9. Connect a wire from the other overload number 1 terminal to overload number 2.

10. Connect a wire from the other number 2 overload terminal to the starter coil.

11. Connect a wire from the other coil terminal to N for 120 volt coils, or L_2 for 230 volt coils.

12. Connect the N.O. (normally open) auxiliary contacts across the start switch (in parallel).

13. Connect the wires from L_1 to L_2 in the disconnect to L_1 and L_2 on the starter.

14. Connect wires from the load-side terminals T_1 and T_2 of the starter to the compressor. (Wire the compressor as a PSC motor.)

15. Ask your instructor to check your wiring.

    Instructor's approval _____

16. Apply power and push the start button. Notice that the starter contacts close and the motor operates. (If the motor will not start, call your instructor.)

17. Push the stop button. The starter's contacts will open and the motor will stop. (If the motor will not stop, call your instructor.)

18. *With your instructor's help,* remove the wire from the start terminal on the compressor. This will cause LRA and should cause the starter overloads to open the circuit. Use an ammeter to monitor the LRA. If the heater and overload are not sized too large, the internal overload in the compressor will open the circuit. Be sure the heaters are correctly sized for the compressor.

19. Turn off all power to the trainer and leave the circuit wired.

TROUBLESHOOTING THE STARTER

Problems in the starter and start-stop switch are easy to locate if you remember that the coil must be energized to close the contacts. Use the following procedure to find problems in the start-stop switch and starter. Check off the steps as you complete them.

1. Ask your instructor to place a problem in your circuit at this time.

2. Apply power and try to make the circuit operate.
   a. If the contacts do not close when the start button is pushed, skip to step 3.
   b. If the contacts close but the motor does not start, skip to step 4.
   c. If the motor starts, check the FLA. If the motor is operating correctly, ask for a new problem.

3. Since the contacts did not close, be sure to check the fuses, and be sure there is power at the disconnect. If power is present, skip to step a. If no power is present, repair the system as needed and return to step 2.
   a. Check for voltage across the coil as you press the pushbutton. If no voltage is present, use the voltage drop method to locate the problem. Make repairs as necessary and return to step 2. (If an overload is open, be sure to check for the cause.)
   b. If voltage is present at the coil but it will not pull in, replace the coil and return to step 2.

4. Since the contacts have closed but the motor did not start, use the voltage drop method to check L_1 and L_2 for voltage to the motor.
   a. If voltage is present at the motor but the motor will not start, check the motor overload. Make repairs as necessary and return to step 2.
   b. If no voltage is present at the motor, check the overload heaters for possible damage. Make repairs as necessary and return to step 2.

When you have found all the problems in the circuit, turn off power to your trainer and remove all wires. Return the wires to their proper place.

This completes Unit 24 on starters. Ask your instructor for the quiz on Unit 24 when you are ready.

Index